TRUE BIRDS

Coline Béry
2015

Copyright 2015 by Coline Béry
All rights reserved, including the right to reproduce this book or portions thereof in any form whatever.
For information address Collection Corde Raide Subsidiary Right Department,
corderaide@orange.fr

Collection Corde Raide can bring author to your live event. For information,
Collection Corde Raide Bureau is 07-68-45-67-13

Interior design
Anne Vanier

Front endpaper : Adrienne Bolland in her Caudron 1920,
Two planes flying in Villa Lugano,
Both courtesy of J.C Boland and F.A.A Archives

Dedication

For Bertrand Piccard and the Solar Impulse

Prologue

The Andes and Adrienne Bolland in 1921

[*This text has been originally published in Spanish for 2A - Ancient Airfields review, in February and April 2015, and in the Gaceta Aeronautica – too. You can read it in English in English Gaceta Aeronautica and in the Adrienne Bolland's Official Website*].

The one who dares going to where nobody ever went and who, by his or her action, widens the field of possible achievements, deserves special attention. The first human (male or female - who knows) to have watched dreamily a bird flying was really a marvelously extraordinary human being.

In the Andes, when it dives from the glaciers, from the highest summits in the down-killing winds, all birds, male or female, hesitate – even the condor. It's normal then that this extraordinary bird should be the symbol of cohesion, strength and courageous union of the people, among other rivals, living in those wide lonely countries.

For eBook version of this book, you will find in the left margins of the text "°°°". Those signs correspond to the 111 illustrations present in the PDF printable.

°°° Google Earth by Anne Vanier.

1921 : At the very beginning of the "Roaring Twenties" in France, the word "Andes" is written in miniscule at the South American page of the Vidal-Lablache Geography Atlas, maps of the time are not fair to the labyrinths and passes, thousands of them entwined among peaks, all of them now measured.

From a plane, over the Atlantic Ocean and 10,500 kilometers away from France, the Andes are evidently the natural frontier dividing, in a dictatorial manner, many people and cultures of South America. But for the people living remote there, the telluric convulsions of the ground under their feet means regularly rebuilding destroyed bridges and crumbling tunnels.

But… what is the relationship with Adrienne Bolland ?

Adrienne Bolland (1895-1975) flew over the Andes by plane on April 1st 1921.

Just before, in 1916, it's not difficult to imagine the collective stupor, the phenomenal fury that the first successful flight over the Andes in a balloon generated. Eight countries are separated by the Andes, two of them – Argentina and Chile – discovered what Eduardo Bradley and Angel María Zuloaga, Argentine pilots, had admired the large immensity of their Andes from the sky (8,000 high): a feeling of admiration invaded the whole South American continent.

Two adventurers had taken off from Santiago: their balloon, pushed by the heavy and warm trade winds of the Pacific, had reached Mendoza, situated on the other side of the frontier, over the Andes – in Argentina. All sorts of geographic possibilities were starting on the 26th of June 1916.

Only five years before Adrienne Bolland left the transatlantic Lutetia in the harbor of Buenos-Aires, four hundred after the « discovery of » Argentina by the Europeans, it was THE event of the year. Of course, the news took some time to reach the limits of the continent, to those few who could read and even the fewer ones who had communication means. Instantaneous transcontinental news did not exist yet. But thanks to the underwater transatlantic cable, installed by the English from 1874, Telegraph[1] was already developed and information could exist. And one day or another, everything became known. Isolated, even if its economic importance was already considered by the same Europeans who had "discovered" and colonized it, South America was always enjoying a privileged status of ancient Eldorado - transformed into a "worldly storehouse". In a spirit of curiosity and open mindedness, South America always welcomed all the technical improvements brought with "heavier than air" – invented mostly by the French Clément Ader - a kind of cousin, parent, brother, friend.

But South America ignored that just as Europe created through her aviation strong phantasmagoric and genealogical links; Europe also exported her worst bad habits.

Sworn enemies, eager rivals, spitefully belligerent forever – and in a state of war since 1914 – the German, English, French and Italian firms stopped their exportations. South America found itself with orders placed in Brazil and Argentina, sometimes paid (as in Chile), but not honored. At the end of the war, after demobilizing of the men, these countries sent their best engineers to Argentina in order to fulfill the orders placed and gain more agreements. External money was necessary for their reconstruction, they were all in common financial weakness - and diplomatic relations depend upon who's brittle and who's strong...

From 1918 in Buenos-Aires, fragile machines were transformed into heavy and strong engines: there were flying guns over the docks of the already extended district of Rio de Janeiro and *Baires*[2] – megalopolis corroded by depression and attacked by anarchists' bombs. But the rich stay rich and the "market" of aviation resists to the chaos. Importations start again.

Builders, groups of instructors, pilots, mechanics and aeronautical material are arriving. The whole world has understood that mastering of the South American sky will be the key to dominate the world and that flying over the Andes is the solution.

Between the different protagonists, peace is ill.

[*Author's note : we must say that from 1921 to 1978 flying over the 1,000 kilometers of the Argentine Pampa, this «treeless place» from Buenos Aires harbor to reach Mendoza at the foot of the Andes, had no interest : the railway, the comfortable Transandin, allowed to make the trip in four days.*]

For the first battle of the air, in 1918, success goes to pilots born on South American ground : Luis Candelaria, an Argentine man of twenty-five years, in a Morane Saulnier of 80-hp^3, is going to fly over the extreme south of the country, where the mountains are less aggressive. He will land on the other side, in Chile, for the first time ever. And on the other side, the first Chilean was Dagoberto Godoy. He was twenty-five years old too. Two young heroes - a new air legend begins.

This achievement was a case of war, for the "flying over the Andes" that the Argentines were expecting. It will impose the beginning of a series of "flying over of the summits in the sky," but in a very nervous atmosphere. Competition begins to know who will be the more powerful, who the lightest, the latest, the quickest between Santiago and Mendoza. All flights need to be efficient – good trade depends on it.

[*Other precision: air roads are not the same at all, from Argentina or from Chile.*]

In the hottest period of this 1920 summer, nobody was expecting a newcomer. French and newcomer, this young person – Adrienne Bolland – has just turned twenty-five – aviation age.

She does not speak a word of Spanish. Who cares about her ignorance of what was Castile and who colonized the country, she feels like crossing swords with the whole air world since she convinced René Caudron to go to Argentine. This young optimistic and free lady has decided to make great flights and a triumphant tour – of course triumphant! – with her two planes.

The industry, the workshop, the firm of her boss, René Caudron (who become more familiar some months ago[4]), needs advertising and orders. The legendary flight over Pekin's Forbidden City, in 1912, is long past and the aircraft manufacturer is being ruined by taxes on war benefits. The heroic raid of Poulet and Benoist (they joined Australia and Paris in a Caudron G.4) is not enough to erase the recent deaths of Jules Védrines and Baroness Delaroche.

René Caudron has thus accepted the request of his most eccentric pilot, after having lectured her about her apparent disappearance when she crossed the Channel on August 25th 1920 ; a success but difficult to communicate upon since Adrienne Bolland was considered lost three times. Three times, papers announced her death on their front page. Three times, Adrienne Bolland denied it, since she was living in a Brussels hotel… after having celebrated with the pilot Lucien Bossoutrot. Because of this raid, the pilot signed a contract of apprentice publicist, and she promised to try to sell two Caudron planes, G.3 type with 80-hp Rhône engine, accompanying her to Argentina. South America needs planes for aeronautical schools and Adrienne Bolland swore not to get into any scandal: only good advertising.

Once arrived, the woman whom French papers already call "Miss Bolland" and her family call "Zizi," thinks she is able to take up any challenge very quickly. Only a few days after her arrival at the luxurious majestic palace, by way of rehabilitation, she sends a telegram to René Caudron that she requires a more powerful plane. She wants to see the Andes – only see them – before flying over – maybe. All the men present on San Isidro airfield make fun of her two "bird cages". But all her hopes vanish when she receives a negative answer from René Caudron. During the twenty-five months of her South American adventure, Adrienne Bolland will receive no powerful plane, nor any money.

Since she has no more hopes, she decides to take any kind of risk, especially when she observes that the mob, watching her aerial acrobatics every weekend above the race-course of Palermo, is bigger and more enthusiastic every time.

The fact that she donates her profits to victims of the Mendoza's earthquake (on December 17th) increases the admiration of the Buenos-Aires populace. Only press people, and the French colony think she is "frivolous". At night, indeed, Señorita Bolland can be seen, unchaperoned and without a hat, in the bad districts near the Rio de la Plata.

In Paris, in Europe, as well as in Argentina, Adrienne Bolland is a young ambitious girl showing her professional talents: on February 26th 1921, she beats the feminine record for altitude 4,850 meters achieved by the Baroness Delaroche; but her altimeter is not certified and her judges cannot homologate her record – even though there are witnesses. For the pilot, enough is enough. Feeling cornered, without money, she decides to leave for Mendoza with one of her two planes.

At first, her only purpose is to see those fantastic Andes she hears so much about since her arrival: However, Antonio Locatelli (1895- 1936), who flew the first postal trip across the Andes, on July 30th 1919, has given conferences at the Jockey Club on Andean crossings. The ten pilots who have already been confronted by this rocky and icy trap were very well prepared for this flying over these summits; but she has no Morane Saulnier type L « Parasol », no Bristol M.1C 110-hp, no Ansaldo SVA 220-hp, even less Breguet Renault with 300-hp. Her fragile sesquiplane dating from 1912 and unboxed only since March 16th 1921, is barded with piano wires - which operate the controls - her wooden cockpit is opened to the wind; and the weary "rototo" (sound of the rotary engine) which spins the 2.5-m diameter propeller, has only eighty horsepower. But, at the Rouen[5] or during the Buc's aerial shows, and during all her trips, in spite of misfires of the engine and defaults of her sesquiplane, she has been elected by the press "Queen of landing with jammed propeller". She's the best with the wind.

Adrienne Bolland is also used to "crashes" but in the French countryside. In the Andes, she will not be able to land or to return and she will never reach the 6,000 meters of height – necessary to avoid the downdrafts – since her limit of height is perhaps 3,000 – 3,200 in icy air.

So… she is no dupe: this raid is a suicide. On top of that, she has her pilot's license only since one year and two months, and she has flown only 40 hours. But who cares: "it's better to be a dead icon going to suicide, than a living pilot but an unknown one for all life. And when I'm scared, I go!"[6]

During her two attempts in Mendoza on March 28th and 31st, what she sees from only 3,000 meters is an experience that fills her head with black thoughts.

She has not been able to sleep for two days, because of fright, when her room's clock in the Grand Hotel (now destroyed) rings four a.m. It's time. The good day to… die ?

Patiently she coats her whole body with grease, covers her bust with layers of newspapers (Los Andes) collected for her for a week, puts on her only turtleneck pullover and over it her best underwear - a heavy silk pajama, which reminds her of her big brother Benoît, her example and human inspiration, who left to live in the far East. In China, Benoît is a boat captain on the Yang-Tseu-Kiang -, then she puts on her manly brown overall and gets out to climbs in her new friends journalist's car. Mendoza airfield (Los Tamarindos, now Plumerillo) is located twenty minutes from the center of this town built in a chessboard pattern.

After four hours of this flight, Chilean aviation students of Lo Espejo (today El Bosque - Escuela de Aviación Capitan Manuel Avalos) see, arising from their sacred mountain, their dream come true… A divine thing, not a pilot. Within a second aged twenty-five, Adrienne Bolland becomes "Goddess of the Andes, symbol of unity and cohesion; the most charming, natural, polite and dignified symbol of good understanding between people. French one.

Exhausted, out of her mind after a "trip in nightmare", in bad shape but alive, Adrienne Bolland is going to embody the eternal feminine gender. Now she represents by herself, all of Europe courage, sent to Chile by god's will and grace; she's all of France and Paris after war's hope, dressed in a beautiful silk pajama: That's the way Adrienne Bolland, and her two legendary Caudron G.3s in South America started a new chapter of Aviation History, and that's how she opened a new page in Women's Rights History.

March 18th 1975 will be the day of the "physical" passing of Adrienne Bolland.

But we are in 2016, the "Miss" was born 1895: we are 120 years after her birth. The President of the Republic of Chili said about it in 1921 : "Blessed is the country which gives life to such rare feminine species[7] !" – Thank you Mr Alessandri.

F-ABGP & F-ABGO

○○○ *Flyer for the G.3 Caudron 1920. Coll. C.Béry.*
○○○ *Landing of a G.3 on Le Crotoy beach. Coll. C.Béry.*
○○○ *The shed of the civil aviation center, brand new at Villa Lugano, with 3 planes of the type Caudron G.3. Revista Aviación N° 12. December 31, 1921. Coll. Dirección de Estudios Históricos de la F.A.A.*

F-ABGP, F-ABGO... and inversely, F-ABGO and F-ABGP[8], are two new codes of the beginning of the Twentieth century, two nearly secret[9] conventions of the brand new technology of "how to fly like a bird" in 1920.

Even without flying over the Andes, the Type G.3 plane created by Gaston and René Caudron - brothers and business men side by side in their firms based in Rue[10], Issy-les-Moulineaux[11] and Lyon-Bron, is, historical speaking, a remarkable plane, but no more than remarkable. The feat of the flight over the Andes by Adrienne Bolland, is going to transform one of those two G.3 Type from remarkable to Legendary within a few hours.

Only one specimen of this particular type achieved this. Now, two similar planes went to South America. And none came back to France. Mystery...

○○○ *1-2- 3 – original plan of Caudron G.3 airframe - Plan of the G.3 dual control. Plan of rudder bar, operated by piano wires.*
○○○ *1- 2 Adrienne Bolland in her G.3, in Buc, October 1920 for "Le Petit Parisien" French newspaper. Her name is painted on the air frame. Coll. C.Béry and J.C. Boland.*
○○○ *1- Caudron Type G.3 (black cross= Italian) used in 1921 at the Civil Aviation School of Villa Lugano, Argentina. Coll. Dirección de Estudios Históricos de la F.A.A.*
○○○ *2 - The Drill Ground – close to Paris -, transformed into an airfield in 1907. Picture taken from Henry Farman Street (the airplane constructor Voisin settles there first, then Caudron) in Issy-les-Moulineaux. Coll. C.Béry.*

F-ABGO and F-ABGP have no names other than their two civil plane registrations, only engraved on their 80-hp engine produced by Rhône firm. F-ABGO and F-ABGB disappeared with everything on board. Lost, with their piano wires and the dress. Yes, there was a cocktail dress and dancing shoes in one of them, and nothing left.

A nothingness, for both of them. As if they had the same story.

They have similar origin; those two planes, those two tubs ("rafiots[12]" in French) sesquiplanes come from the Issy-les-Moulineaux's workshop. They have, identically, apart from their G.3[13]Type, the same initial letters, with the same hyphen, separating the F (for France) from the rest of the required registration. The two "twins" having GO and GP, letters following logically in the registers, are assembled successively in the Hexagon[14] to promote their builder and find buyers in South America.

°°° *1-2- Advertising for Caudron, and their two shed-firms in Issy-les-Moulineaux, 1920 – Coll. C.Béry.*

The Caudron's Firm[15], in a delicate situation money-wise since the end of the First World War[16], has already sold G.3 types to Argentina; but René Caudron ignores that the two "prehistoric" machines he is sending by ship, will be symbolic of two episodes, a bright one and another one, scorching, for international aviation.

Who could have imagined what was going to happen In South America ? Caudron ignores that he will also stay linked to the brand new and ambivalent fate of a young pilot already know overseas as "La Aviadora francesa Mlle Bolland"[17].

°°° *1-The Lutetia Transatlantic ship, boarding. 2 – Cabin 1st class on the Lutetia. 3 – View of a pier of Buenos-Aires's harbor in 1920. Coll. C.Béry.*
°°° *1- G.3 model kept at the Dirección de Estudios Históricos de la F.A.A. – Picture Anne Vanier. 2 – G.3 showing by Carlos R.Boisen. The stamping of the philatelic department of Argentine postal service has been realized in 2011, to commemorate the 90th birthday of the crossing of Adrienne Bolland, on a request of the colonel G. Pavlovcic and C.Béry.*

The two planes, with identical airframes, yellow[18] and without any trace of black letters – we know it today – leave Bordeaux's harbor on Saturday, December 4th 1920.

But instead of going back inside the store rooms of the Transatlantic Lutetia[19], at the Rio de Janeiro stop – they stay alongside the quay: "General strike!" announces the Pacha to this special passenger - not yet "pilot" but already a real "young star" of futurist aviation technology. In fact, the flight of those "heavier than air" is also celebrating twenty-five years of existence. The sesquiplane Caudron G.3 was born in 1912, its history dating only eight years, but the pilot of the two planes just turned twenty-five on November of 1920.

Apart from having the same age as her new "job", she can also boast of being the first woman to have crossed the Channel by plane (Wednesday August 25th 1920) like Bleriot, from France to England[20]; the first one to perform acrobatics figures[21] at a French aviation meet (Rouen, April 1920); the first one to be sponsored by an airplanes constructor who also employs her for flying planes to purchasers (since March 1920).

°°° *View on the Corcovado of Rio de Janeiro 1920. Coll. C.Béry.*

General strike is a problem... But Rio dockers are in anger because of their extra work never paid, and fanatic anarchists take this opportunity to blow themselves up. Rio quay is a perfect target. Consequently, the boxes must be taken out, verified, checked, and inspected. When the Transatlantic leaves for Buenos-Aires, she carries only passengers.

The two G.3 F-ABGO and F-ABGP – promotion tools – will stay two weeks in Rio. Their pilot will disembark "bare handed" on Thursday, December 23rd in Buenos-Aires. "What a diplomatic restraint for defending our colors in this nation, so foreign country", complains the French consulate and official expatriates at the Buenos-Aires air club, minds not disposed at all to woman aviators - women *too much liberated*, as everyone knows.

She doesn't care. The two G.3s will soon be ashore in Buenos-Aires.

°°° *May Avenue between the Rio de La Plata and Congress in the back. Majestic hotel is show in red. Coll. C.Béry.*
°°° *1- The Majestic Hotel 2011 - 2 - From one of the balconies of Majestic Hotel 1920, view over May Avenue and Congress – in the ground, Buenos-Aires light house. Coll. C.Béry.*

From the Majestic Hotel, Buenos-Aires Ritz, Adrienne Bolland announces on January 2nd 1921[22], that her two planes arrived and that, as soon as they can be reassembled, the promotion and advertising tour is going to start. Advertising/promotion in aviation is called Air Propaganda, in French[23]. Advertising claims indicate that Adrienne Bolland had already worked at this with a G.3 during three days, in Buc (8-10 October). Fonck, Romanet, Casale, Boussoutrot, Vuillemin et Fronval, the Heroes from Buc's meeting are doing their boss's propaganda too, in the beginning of 1921, when Bolland's two G.3s join effectively the Breguet 14, SPAD S.13, D.H.16, D.H.6, Avro 504, Nieuport 28, Curtiss JN and Curtiss Oriole, side by side on the San Isidro field (sixty hectares transformed in an airfield[24]), with the French Aviation Mission planes, settled since the River Plate Aviation Co., managed by the powerful and imposing Major Kingsley[25] - English man.

Caudron's name, French one, is well known and admired in San Isidro, and the accreditations signed for her by Mr Flandin[26], Minister ; by her boss, known around the world since the sales of his G.3 in China[27] ; and by her friend Auguste Maïcon, founder and manager of his "Compagnie Méditerranéenne de Transports Aériens" based in Nice[28], mean that Adrienne Bolland is in order.

°°° *1 - René Duperrier in Mendoza, in front of the monument built in commemoration of the Argentine army, just before the flying over the Andes of the G.3a. . 2 - 3 - 4 - At Buenos-Aires station, after the pilot's feat. 5 - In front of G.3b and members of the Air club of Buenos-Aires.*

Her mechanic, René Duperrier, self-confident and cold-blooded young man, does not only repair planes; pictures always show him next to Adrienne Bolland. He is her professional partner, indeed, but also her confidential agent, the one who holds her life, their lives in his hands, since he is often Adrienne Bolland's passenger. He takes in charge communication, he is a bulwark against misogyny, and finally her only accomplice and "adventures" friend. All the professional appointments of Adrienne Bolland are handled by René Duperrier, but before he must direct the disembarking of boxes and the logistics of their settlement in the sheds. In San Isidro, the mechanic easily remarks that there is a big difference between the powerful planes of other pilots and, their two sesquiplanes built in sequence back in 1912.

°°° *Florencio Martinez de Hoz, Président of the Buenos-Aires Air Club, Argentina 1921. Coll. Estudios Históricos de la F.A.A.*
°°° *Advertising of Bessonneau's firm, l'Illustration 1916. Coll. C.Béry.*

The French Aviation Mission is composed of Captain Rémy Vuillemin and his mechanic Caussé, of Lieutenants Jean Guichard, Fernand Prieur and Charles Hildebrand, of sergeants' major Daniel Clerté, Adrien Bedrignans, Marcel de Saint-Sulpice, and sergeant Georges Combes. A never-ending coming and going of pilots, civilians, soldiers, business men, eminent persons and international mechanics takes place in San Isidro, following the rythm of Transatlantic crossings.

The work takes place in three fragile, temporary sheds, Bessonneau type (some month later, they will be seen while being assembled at Lugano's Air Club). Meanwhile, Major Kingsley's River Plate Aviation Co. is sheltered in a huge solid shed, with a metal structure, wooden walls and sheet roofs, divided in fifteen meters portions.

°°° *Building of the Civil School Aviation in Villa Lugano, 1921. Coll. Estudios Históricos de la F.A.A.*

Transportation of passengers by one or the other Mission and new Air line involves a promising financial advantage. On the regular lines openings in the wide Argentina territory, returning customers become strong economic and political partners. Future air fields are often located on meadows belonging to important land owners, careful to bring modernity of the world in their empty lives and greedy to "shine" while learning to dominate a "heavier than air"[29].

Building and selling planes, maintaining, transforming them, and training civilians or soldiers are competitive activities which generate regular tensions in the different Argentine locations.

Marcel Paillette[30], a French pioneer, and well known military instructor working at the Military Aviation School, uses six G.3s for reconnaissance flights and aerial photography. The rollover accident of his twin-engine G.4 on December 7th 1920 does not discredit Caudron. Marcel Paillette, who survived worse before, is not surprised by this young neophyte girl with her temperament totally dedicated to practical things and piloting with panache and talent. She obtains the audience's approbation with her rolls and loops: well done Miss! but the engine's high resistance though is more surprising.

Strengthened by the impression she gives to her audience, which grows more and more numerous every day, Adrienne Bolland announces she wants to "see nearer" the Andes – once there, her Caudron and herself will decide.

Newspapers break loose and tell her "Miss, 'our' Andes are not chocolate mountains", and she answers "but my Caudron is not marshmallow": which tells a lot about what "her[31]" Caudron means to her. Her faith entertains some, her will gets on the nerves of others; all of them agree, though, that with the criminal attempts of anarchists going on in the city, this acrobat who fills the race course without being paid has no doubt "l'esprit d'à propos".

°°° The Andes next to Uspallata's plain in 2011 and in 1920. *Coll. C.Béry.*

With or without audacity, flying over the Andes means either jackpot or death. The pilots'[32] brand new Heroes are celebrities, and their fame around the world halo them. Unforgettable men now, they all give conferences in the most selective clubs of North and South America. Adrienne Bolland meets Vicente Almandos Almonacid and Antonio Locatelli[33], whose success warms up the walls of the Jockey Club, they all say she will never be victorious, and the truth is that she does not imagine being. Only to TRY. But even trying is a suicide. Nevertheless, if she dares listening to her instinct and if she manages to reach Chile alive, she will be the ninth person to fly over those deadly mountains and, moreover, she will be the first woman in the world to see the Andes from the sky.

Glory, wealth and well deserved rest are linked to this flight over the Andes.

From Thursday December 23rd to Saturday March 19th: during those nearly three month in Buenos-Aires, she is going to take advantage of a still summer season (seasons are reversed: deep winter in Paris[34] is, in Argentina, the end of a humid and stifling summer). She is also going to take advantage of every event - even the earthquake in Mendoza on December 17th - to show the talents of "her" Caudron and offer to inhabitants of Buenos-Aires and the Rio – Porteños -, notable exhibitions.

Aerial acrobatics - not yet called "Voltige" - speed, risk, and fear are her favorite weapons. The audience loves it.

°°° *Palermo air course where Adrienne Bolland offers sessions of acrobatics in her G.3 to Argentine audience.*

All the pilots who have flown in a G.3 Caudron describe a terrifying but ecstatic experiment.

Your mouth absorbing the thick fluid of air; receiving gallons of castor oil in your face; wiping your goggles while trying to see what comes ahead[35]; managing the rudder bar and maintaining the stick, require intense energy, cold blood, good resistance to cold, and a level head since the rotary engine, not powerful, has misfires.

When the engine "misses" a beat, the propeller misses one too, and stops. The plane is supposed to soar a little, but in time it crashes.

°°° *Caudron G.3 Model. Coll F.A.A.*

A sure hand and instinctive sense of the wind, endurance of a highly trained sportsman and pugnacity are the qualities required with a G.3 Type Caudron, whose pilot flies through the air without any protection of the torso, with arms resting on the edges of the seat. A hundred meters of piano wire, extend to every strategic point of the plane for "gauchir", in French, the warping of wing surfaces to control roll. Considering the absence of ailerons (rare in the day), the plane throbs tremendously. And, when it lands – within sixty meters, a very short distance - the rubber shocks absorbers of the wheels don't help a lot to absorbs the shocks of the ground.

The fabric over the seat and the wings are major problems in case of icy rain or hail storms. In a high altitude and in case of extreme cold, the Caudron G.3 is a very feeble bird.

°°° *Palermo race course where Adrienne Bolland is a star every weekend in Buenos-Aires.*

But none of her Caudron's defaults are an obstacle to her pilot's ambition. Indeed, on Saturday February 25th at 8:15

a.m., René Duperrier is with her when they reach an altitude of 4,850 meters around 12 o'clock. This record, qualified by the Argentine newspaper of "a sports feat opening the future for women to new feats" is not homologated: the altimeter is not certified. But this particular G.3 has remained airworthy and, in front of new disparagement of the French colony, Adrienne Bolland decides to advance the limits of her planes. Acrobatic programs and regular exhibitions at Buenos-Aires race course – a smart and popular meeting place for air meets – are compelling her to a decision: she must live this mad, and so exciting, thing[36]... Trying to cross the Andes will be setting "her" mark (in French "marquer le coup").

René Caudron refuses to send her a more powerful plane to perform this "madness". The pilot is alone with her conscience, "lightened", she recalls in 1951[37], by the mysterious words of a woman visiting her, and offering her Virgin Mary medals to protect her during her flight. It is announced in the international media[38] on Wednesday March 18th, she has just one of the G.3*a* (we name "him" G.3*a*) boxed and transferred on the Pacifico[39] Line train, destination Mendoza (1,000 kilometers west of Buenos-Aires).

The G.3*b*, twin of G.3*a*, stays in Buenos-Aires: Choice made.

ooo *The Retiro Station in Buenos-Aires, where the transandine Pacifico Line Railway leaves, going to the capital to Valparaiso, Chile. Coll. C.Béry.*

Condor of Stones

°°° *The transandine tank engine in the Andes, with ice-drilling rostrum. Archivos Generales de Mendoza.*
°°° *The Andes before arriving in Mendoza.*

When the train arrives on Saturday, March 20th 1921, the station is overcrowded ; an anxious and curious mob follow the pilot, her mechanic, and the officials[40] welcoming and taking them to the Grand Hotel located on España Avenue between the streets Necochea and Las Heras (today it has been destroyed).

°°° *Railway station of Mendoza 1904. Archives Générales de Mendoza.*

Mendoza is a city at 1.000 meters altitude, built at the feet of the "obstacle". For all time, and for fascination to all, this monster is nearby. Headaches and anxiety invade the pilot. Time goes by quickly indeed, here it is already autumn, and once more time, Adrienne Bolland must wait for the G.3*a* to be reassembled. She will help of course - she always helps and questions Duperrier to understand "how her G.3 works", so she helps and consults maps and talks with people living in the mountains. She watches the mountains, stopping sunlight at 3 p.m. - As she faces south, the horizon is all closed on her right.

Her only activities are shooting sessions and horseback trips – horses lent by the military in Tamarindos[41] (the military airfield is located in the desert, five kilometers outside the city). This airfield is named Tamarindos, because it has been planted with palms imported by 16th Century settlers, and irrigated by alluvial deposits.

°°° *Adrienne Bolland, mechanics Crochard - on her right, and René Duperrier ; Ernesto Escobar Bavio, for El Mercurio, and don César Sagredo director of Los Andes.*

She has decided which air route she will follow: the northern one - the most dangerous; the one that requires flying as fast as possible and an altitude of 6,000 meters over the downdrafts. Power and speed? Rene Duperrier - and everyone present at the airfield during its announcement - consider it as a bad joke. She wants suicide?

"It's a conscious unconsciousness, but I'll do it[42]", so, Duperrier modifies the G.3*a*: he installs another fuel tank on the forward passenger seat. The craft's maximum flight time is doubled and is now eight hours.

°°° *Mendoza city, actual view.*

After this enhanced autonomy, the Caudron must still reach the altitude needed. Adrienne Bolland is warned: she will have to be as "light" as possible. Other pilots interviewed in Buenos-Aires concur on this topic. Benjamin Matienzo, even "lightened" and with a powerful aircraft much superior to her plane G.3, crashed. If Matienzo crashed, it was not due to his plane, or to this tunnel, but because of his lack of confidence. Doubts overwhelmed this military pilot. The solution ? Flying at the highest possible altitude.

Without sophisticated flight instruments, without oxygen bottle, without closed or protected airframe, death is sure. And the Caudron will have to withstand "inhuman" temperatures for a plane. And for the pilot, at 3,000 meters, the air will be an unbreathable pea-soup. In conclusion, for her lungs and the plane's carburetors, their fate is linked – the icy air will interfere with propulsion and kill the pilot in a few hours. At this altitude the temperature is – 20 Celsius.

°°° *The Andes seen from a Cessna biplane, 6,000 high all rights reserved Philippe Rolet.*

Moreover, she will need to remember to take off her shoes! And, if there is condor on her way, she can say goodbye to the propeller. Just in case, she will take a dagger... and a spare magneto providing the engine's ignition. At the first trials in Mendoza, Duperrier announces that they don't have new magneto. Buenos-Aires must send it by train... before killing herself, she will wait few days.

°°° *Detail of the manual of the Rhône engine 80 hp for the Tpe G.3 Caudron, and location of the magneto. Coll. C.Béry.*

The first trial of the flight over the Andes starts at 4:15 p.m. on Monday, March 28th.

It should have started two hours earlier, unfortunately when the pilot arrives in the Tamarindos, the plane is taken out from under its tarpaulin; it is surrounded, sanctified, touched and prayed for by hundreds of sandy hands; it is blessed by Indians living in the first valleys of the sacred mountains: this flying object unidentified in their culture, but surely magic, could questioning the sky, intercede for them, and maybe stopping earthquakes?

"Kidnapped" in this feverish celebration for two hours the trial start too late, then, without a good quality ignition and good weather forecast, Adrienne Boland realizes that the G.3*a* is not likely to go very far. The day nearly over adds to all winds : - from the Pacific, - from Patagonia 2,700 kilometers south, - from the Antarctic, coming up frozen, - from the Brazilian steamy jungle, 3,000 kilometers away, becomes storming hails.

Cyclonic downdrafts, spouting up and fusing against firmament, transform the Uspallata valley in a wall which stops her. At kilometer 81 of the railway – this railway counts 1,439[43] kilometers and 250 of them are between Mendoza and Chile – she must go back.

°°° *The small Andes, already difficult to fly over in a G.3 because of downdrafts. On the left, entrance in the mountain location "Los Potrerillos".*

Back to earth, the Caudron shows its frozen shrouds and burning piano wires to René Duperrier; its oil is densified and the surfaces marked by being tossed around. Another « gliding plane » trial takes place on March 31st. Three days of anxiety and no sleep for the pilot. The G.3*a* Caudron starts its slow upward spirals at 7 a.m., but it is seen once more coming back. Wheels on the stony field at 9:10 a.m, its metallic parts and wings are, once again, frozen. "No more flying today!," announces the morning press. "Will she make it[44]? - no, she won't," wonders everyone in the town, invaded by people betting illegally.

°°° *1 – G.3a under its temporary tent on the Tamarindos camp, Mendoza. Private coll. 2 – Puente del Inca, The Inca's bridge, a historical natural side 1904. Archivos de Mendoza.*

Maybe it's this special night that will give René Duperrier the habit of sleeping next to the plane. He knows it by heart now and cares for it as well as for Adrienne Bolland. He fears that someone of ill will might come and ruin his efforts to maintain alive this "wood's pile" – "tas de bois" in French. And, on the right day, he himself comes and knocks at Bolland's hotel room at four a.m. on April 1st. The G.3*a* is ok, even then the weather forecast is good – no clouds, sky's open. She is the one to say she agrees on a new trial.

The trip that the two are going to make is watched, even spied, by employees of the different telegraph services and by several people in their vicinity. For ten days already, messages have gone out to the Provinces, all Argentina, all Chile, as far as Rio de Janeiro. A plane, a giant beastly flying object living under the strange name and code of G.3

Type Caudron Rhône engine 80-hp, piloted by a woman "with bird's face and looking like a Parisian *Gavroche*," might soon take off from Mendoza…"

But only the telegraph employees are warned on April 1st, that a G.3*a* with a lightly clad woman as a pilot actually takes off towards Los Potrerillos, without any praying Indians around to offer their blessing. La "Señorita Bolland" is alone and travelling light (she has pushed a bag under her seat with an evening dress and dancing shoes in case of a miracle). She is undertaking the air adventure of the century.

°°° *Taking off from Los Tamarindos, air field of Mendoza, Argentina, for a 2nd trial in 1921. Coll. Archives from the Historical Institute of the Argentina's Aerial Force.*

The plane's wheels forsake Argentine ground at 6:32 a.m. after a few minutes of stress - at first, the plane's engine had stalled, then ran properly. Far from the dozen of frozen people crying[45] in Mendoza at 6:50 a.m., G.3*a* can be seen entering the Andes.

Wind gusts at the valley entrance grow stronger in the middle of it. Pushed, the G.3*a* is located at "Los Potrerillos" – wide lake with 3,000 meters under its airframe. At 7:35[46] the plane is hurled, shaken violently from one side to another. The axial screw[47] and its rudder bar – and also two visual landmarks, the rails of the Transandin and the river in the valley - become Bolland's most loyal flight instruments[48].

The pilot reaches kilometer 81 of the railway at 7:46 - 7:52, where she had been stopped the day before.

7:58 a.m., Uspallata is a desert valley at this point of the Andes which is forty kilometers wide and three hundred long: a sea of multicolored, mixed, agglomerated rocks drifting from the summits and mixed on the ground since the birth of the mountains. There, facing each other, the winds provoke "stationary moments", of fifteen minutes.

During fifteen minutes, G.3*a* creeks and cracks, its propeller fights in a totally frozen atmosphere. From 8 to 8:21 a.m. (kms 103-108) then between 8:40 and 8:55, the plane is caught and captured by the air.

The pilot can no longer feel her frozen arms. The plane resists until 9:10 reaching the "Puente del Inca" (a stony rock place hollowed by running waters doubled by another man-built bridge), then a white triangle, the 7,000 meters Aconcagua, appears suddenly.

Drops of blood stain the fuselage; the pilot no longer has any feeling in her nose; incredibly, two meters ahead of her iced face, the black roundel of the propeller still functions… The engine regularly beats time, with the wheels 3,200 meters above ground in an emptiness to which their mass seems glued.

°°° *GPS points of the flying time – 9.20-9.30 am : 32°49'09.14'' S – 69°56'23.51''O. The famous « oyster shaped lake » in 1921 was on the right. Ever since, the river deposits moved it in a marshy laguna. Google Earth view, Anne Vanier.*

After more than two and a half hours of flight, the G.3*a* reaches a fork: a large valley opens with two others valleys from one foot to another of the Aconcagua: the white and beautiful point of view hypnotized the pilot.

From 9:20 to 9:30 she slows down, loses carefully altitude and then turns gently over a place called Los Horcones. The risk of stalling is foremost in her mind. The wings of the Caudron flap irregularly now, due to the hot gusts arising from the ground, hotter than the air at altitude. She hesitates – a long time…

Where she is, ten minutes is an eternity and she had doubts: (Should she go right into this very attractive and lighted valley[49] - but leading to another invisible valley, or go straight ahead?) Then, when she bends over to see where she is going, the rubber of her goggles explodes !

°°° *Entrance of the tunnel of Transandine railway disappears in the mountain en 1904. Collection Archives of Mendoza.*

Without goggles, she notices that the railway disappears into the mountains. A black hole. Here is the tunnel where Matienzo has been found dead.

On her right hand, the valley bend towards the north, to a new, awful labyrinth of black valleys and other summits, but not at all towards the Pacific; and ten minutes are necessary for the Caudron to reach, on the left flank of the mountain, the location called "El Cristo Redendor."

°°° *The Cristo Redendor of the Andes in 1904.seven meters high bronze, crossing road, frontier of passed centuries, today sacred place of pilgrimage, 3,832 m high between Argentine and Chile. Collection Archives of Mendoza.*

9:38 : The Caudron flies dangerously with its allowable altitude, she can see the fossil crest where the millenary pilgrimage path reaches a double path, on one side climbing up in "S" turns from Argentine, and on the other side very steep too, going down to Chile. Two funnels, glued together, an hourglass crossed by the G.3*a* at 9:45 am, this neck called "The reversed Chair" - "La Chaise renversée" in French, thanks to the Venturi effect[50], is the only means of crossing, the unique pass for a plane whose altitude is limited to 4,200 meters.

Once beyond the "S" called "Snails" ("Los Caracoles" in Spanish), the river waters flow the opposite way, clear and descending towards the plain limited by the blue edge of the Pacific.

°°° *The telegraph posts announces at 9.45 am that the G.3a has flown over the Cristo Redendor comb – at the crossing of the frontier – yellow line and the blue line – way of the G.3a. The plane has been able to fly in the hollow located just after the Cristo Redendor. Picture, A. Vanier.*

At 10:10, after having passed along the west side of the Andes, with the Pacific seen on the right, and searching for Santiago[51], farther south, the Type Caudron G.3*a* finally lands, propeller stopped and engine off, in the sand of the Military Aviation School of Chile called «Lo Espejo» (today El Bosque - Escuela de Aviación Capitán Manuel Avalos Prado).

Within the four hours and fifteen minutes of flight, the G.3a didn't know misfire.

°°° *The flying school Lo Espejo/ El Bosque, on 1920, as Adrienne Bolland discovers it, arriving North of Santiago. Archives from the Military Flying School of Santiago.*

Friday April 1st of 1921, since now, this day is the day of Adrienne Bolland's flight across the Andes.

On this 1st April 1921 where the G.3a Caudron is united to Chile, the airfield and school have been existing for eight years[52]. In 1921, both depend on Government Alessandri[53] and are managed by General Contreras Sotomayor. Chile has recurrent problems at the Peruvian frontier on the north and the policy of buying planes and instructing student pilots, as determined by Major Scott, is exemplary of this fear.

Originally, the school planes were French ([54]two Blériot 80-hp, three Blériot 50-hp, one Deperdussin 70-hp, one Voisin Renault 70-hp, two school Blériot 35-hp, one Sanchez Beza 80-hp, one Morane-Saulnier 80-hp second hand to Argentina in 1917, and two penguins[55]),...

°°° *The aerial fleet of the Piloting School of Santiago in 1913. Coll. Escuela de Aviación de la Fuerza Aérea de Chile Capitán Manuel Avalos Prado.*

... but on July 26th 1918 – date of birth of the first sheds in 1921 - nineteen De Havilland, six Scout S.E.5 and twelve Bristol[56] were added. English presence, at the arrival of the Type G.3 Caudron 1912 piloted by Adrienne Bolland, can be explained by the impossibility of the government of Her Majesty to forward to Chile the orders of flying machines even if they were already paid. With the end of European conflict in 1918, England gave very sophisticated material to Chile as compensation for the country's "constraint." Furthermore, the first "Blitzkrieg" made the French manufacturers' "Demoiselle" and "Antoinette" aircraft obsolete.

Planes become flying registrations, "reasoned" firearms.

In 1920, in the sheds of the Chilean site, there are twenty de Havilland D.H.9 240-hp; eight R.A.F. SE.5a Scout 220-hp; twelve Bristol M.1C 110-hp and two Avro 504. For Navy there are three hydroplanes Sopwith Baby 130-hp; two Felixtowe F.2A 345-hp; six Short 184 260-hp and two Avro 504K 130-hp^{57}.

The *G.3a* Caudron piloted by Adrienne Bolland, when landing at 11:10 a.m. (the hour written in all articles in France given in Argentine time!), comes as a visual and technical bomb. In the competition towards industrial technology, the pilot is not a French princess dressed with lace that Chile and Santiago inhabitants expect. They take from her plane a frozen infant, and breaking her flight harness, they name her "Goddess of the Andes". A Goddess managing a "wood pile" – another nickname of this plane.

°°° *Reconstitution of the « surprise » of the arrival of the Caudron in Santiago, filmed by the Fox Studios. National Archives of Chili.*
°°° *1 2 3 4- Adrienne Bolland, French woman from French Loiret's district, here in Santiago on April 1st 1921. National Archives of Chili.*

Added to the number of flights of the pilot (now a few hundred more, all in South America) are the four hours of a bewildering swiftness for the time. Four hours that makes her "avia-trix" – the equal of birds.

Her stay in the Chile capital is short (Friday April 1st - Tuesday April 5th 1921).

Meanwhile the Caudron is being thoroughly maintained, as a survivor, and her pilot consents to a reconstitution of "their" landing, photographed from all angles and filmed in the afternoon sunshine by a representative of the brand new Fox Hollywood studios. The Santiaguinos approach their new Goddess on Sunday April 3rd, when she takes part – dressed with her famous silk pajama - to a national commemoration: the baptism of the De Havilland newly acquired by the State.

« El Ferroviario » and the G.3*a* Caudron are the glories of the day. Captain Diego Aracena (future hero of the Santiago-Rio de Janeiro raid in 1922) maneuvers the beautiful De Havilland with talent; the audience, including ministers, diplomats, church authorities, and of course pilots, are in admiration of Adrienne Bolland and her G.3, already adopted, and carefully kept on the location by Chile.

The G.3*a* is particularly studied by Clodomiro Figueroa (civil pilot, owner of a Nieuport), who will soon pilot the "bird cage" himself its owner pilot will have left for Argentina.

°°° *Pirque Railway Station, now destroyed. Web site « Nuestro Chile ».*

Train is safer than plane in crossing the Andes. It becomes her unique way of travelling to Buenos-Aires. (While in Mendoza, all of the taxi drivers give her free rides.) In Mendoza and Buenos-Aires, the atmosphere is crazy: A series of anarchist attempts gets on the nerves of the aviators who surround Adrienne Bolland. Hundreds of thousands people massed around the official sports car where she has to stay standing up on the back seat to be seen and admired as "the lady without a hat", a French and young lady – nearly naked in public – and an exceptional pilot. The two of them are protected by René Duperrier and some "military bodyguards", new aviators and a famous one. Adrienne Bolland overcomes them, but the expression of her face mingles happiness with anxiety…

No plane is available to join the ex-French service Men Club, near Congress; the planes are in the sky: a flight of biplanes and monoplanes come to greet the "winner" and fly over the crowded avenues of the center of Buenos-Aires. The festivities start on Saturday April 9th 1921

°°° *Some international covers, after the flying over the Andes of Adrienne Bolland in 1921.*

End of the Condor ?

°°° *The G.3a Type Caudron in front of the Flying School of Santiago on April 1st 1921. Revista Zig-Zag, National Archives of Santiago, Chili.*

Now considered as a legitimate object of worship, the G.3*a* residing in Santiago is now going to awaken desires of numerous pilots and business men.

Adrienne Bolland is without money for three months – René Caudron did not send her any money since her arrival in Buenos-Aires. She tells a mechanic (the name of Fernando Mutt appears in the archives) to advertise and sell the G.3*a* by auction. A sum of five thousand pesos is mentioned in the papers. Diogenes Valenzuela (his name is in the file but has not been verified) buys it and wants to give it to the National Aviation Society of his country.

But myths always have different versions… version II is that Adrienne Bolland sold it directly to Aladino Azzari, a sportsman, future famous race driver, for the same sum.

°°° *Several Chilean publications concern the history of the G.3a residing on their ground, and several photos represent the field of Lo Espejo in 1921. 1- G.3a on the field of Lo Espejo. Coll F.A.Ch. 2 – The pilot and aviator Clodomiro Figueroa in front of the G.3a coming out his shed, unknown date. Coll. F.A.Ch.*

According to the historians of Aerial Force, Aladino Azzari learns to fly in the mythic Caudron; he lends it gladly to his friends, such as Clodomiro Figueroa. The aviator Mario Pozzati takes several photographers for flights and then one day, the Caudron-suffers from a failure of petrol, it crashes when landing - and this is the end of the G.3*a*.

°°° *Aerial photography of 1922 : San Bernardino district in Santiago, picture taken from the G.3a. Coll. I.I.H.A.CH Institute of Historical Investigations of Aerial Force of Chile.*

Version III is already identical to II; but in this one that Mario Pozzati buys it in June 1923 (thus after the crash) and establishes a National Airline Company. The G.3*a* becomes an advertising plane, is used for first flights and as training plane (as it will be in France until 1928, since more than 17,000 of the world's pilots will receive their licenses with the Type G.3 Caudron).

The plane, in version III, is really a victim of an accident. It is carried, on a still unknown date and in a miserable condition as we imagine, to the flying school El Bosque, where in spite of their efforts, the mechanics fail to repair it.

It can appear surprising, when one knows the ability of the talented hands of those schools to reinvent flying machines they receive regularly. An aviation shed, at this time, is the place where a mix of parts and mechanisms is possible. Adaptation of machines to their environment is made on the spot during assembling. Many things change according to needs and following the seasons. It is sure that the G.3*a* has provided screws, bolts, strands, piano wires, and engine to other machines[58].

Soon forgotten nevertheless, various parts of the plane are put in a corner of the shed then transferred later to the "Arts and Crafts School" of Santiago, where they are used during "the practical course of mechanic." Then a fire takes place.

The warm oily flames of the shed destroyed what remained of the legendary G.3*a* -which had survived the extreme cold on the roof of the South American world…

Air, fire, ice, earth: the elements have overtaken and finally destroyed this symbol of early aviation. Yet, this ultimate disappearance of Version III does not quite satisfy the Chilean Air Force official papers, but it had to disappear, otherwise, it would have been seen in the press during a commemoration. So, this is still a mystery.

One thing is sure, the legendary flying machine ended its life in Chile.

And what of Adrienne Bolland's cocktail dress taken aboard April 1st ? No one ever found it after her landing in Chile. Maybe a student pilot will have wanted to keep for himself this fragile symbol of international aviation?

Toucan of Sands

°°° *Adrienne Bolland standing in front of her sportscar, drowned in the mob of 100,000 men in the Buenos-Aires street on Saturday April 9th. Coll. Archives of the Nation Argentina.*
°°° *Adrienne Bolland and the G.3b, Buenos-Aires.*
°°° *And the day of the famous picture Coll. Historical Institute of the F.A.A. and photo in the Aéro-Club of Buenos-Aires since 1921.*

Now the G.3*b* can live.
Staying in the shed of San Isidro since past Saturday March 20th, it was waiting for its chance.

The pictures taken a few days with the return of its pilot show the Caudron painted with the letters MADO. on the front of its steel cowling. Usually, inscriptions can be seen on the airframe. But it is difficult to take black paint off the seat's material. Thus it is the owner of the plane who decides "the name code" to be written in big letters to be seen from afar.
In April 1921, writing this name code on the steel part of the G.3*b* (which was meant to find a buyer, as its twin brother) allowed it to be erased more easily after the sale.
MADO.
A mystery for more than ninety years.
In 2015, an inscription also written on the cowling of another engine, dating from 1922, will allow solution of the mystery. BANDEIRANTE.
Anésia Pinheiro Machado is eighteen years old and the daughter of wealthy Brazilian planters originally from Portugal. She's going to buy a plane and will have the word BANDEIRANTE inscribed as homage to her adventurer ancestors, who came in the 17th century to live upon mineral and human resources of Brazil. *Bandeirante*: or maybe Anesia Pinheiro Machado is talking about herself? It's also

the name of a place, far from Rio de Janeiro and still wild, in 1922.

°°° *Anésia Pinheiro Machado in 1922 - Wikipédia*

Adrienne Bolland is twenty-five years in April 1921. She is French and famous within one day. Her nerve and will evolve from "unbearable whims" to "wonderful feats" in only one day too. She has known the world war in France, she loves freedom, fashion, the French graphic spirit, and its musical version, jazz, jerk; she loves partying, speed and car or horse racing, she needs anything that dances and moves frantically at the beginning of the "Swinging years".
Even at the other end of the world, far from Paris, she is "in" - and herself: a friendly and cheerful young person used to giving nicknames[59] to the ones she loves. And she is superstitious.
Her plane is her work tool, and the expression of her real being. Her superstition comes from her past. It was born, so to say, on her birth date. November 25 is also the birth date of her beloved father. This superstition became permanent, when in October 1919, to forget her disappointments (she bet all her money on a race course not to be dependent on her widowed mother) an unknown man, met in a Paris night club, speaks to her of aviation as a possible job. Fortune, liberty of action and glory reign: and Aviation is still a world without women. A brand new country. It is above all, the universe of danger, of fear – and chronic superstition. The flying school of the Caudron which trained her at Le Crotoy[60] notices that she has refused attribution of N°13 for her flying license[61]. When this unknown woman visits her just before she leaves Buenos-Aires, and when she gave her the clinging medals of Virgin Mary, blessed, she surely offered magical charms. Adrienne Bolland put these medals in her pocket before her flight. They surely have protected her Caudron.

The young pilot sees a logic in all that miraculous events. The G.3*b* is the ideal means of telling the whole world that she thanks the holy virgin, thanks medals, thanks unknown visitors, thanks the bets in Paris,... these letters are painted for the press to publish them (painted on the airframe, they would not be visible on the official pictures). But abbreviating MADOnna to MADO. (and not MAD, which is negative in English) on the plane's muzzle, with its strange "." is the most efficient solution – and the most inconceivable too, since everybody knows in Buenos-Aires that the pilot's behavior is in total contradiction with the moral principles of the Catholic church. Superstition beyond suspicion.

Of course, the big black letters are going to be printed in all the papers, even the who's who of this year 1921. The echoes of this new fashion will reach Brazil first. As seen before, the MADO's model is going to inspire Anesia Pineihro Machado.

In Brazil, as everywhere else in the world, trendsetting youth likes to show off its brand new secret codes.

The G.3*b* MADO., though not responsible for of the legendary flight of its twin, will nevertheless receive the honors of a whole country and continent from Sunday April 10th. On that day, the forty most influential people of Argentina welcome its pilot, without her mechanic friend, in their Holy of Holies Congress.

°°° *Adrienne Bolland received in the Congress.*

Speeches, parties, banquets, celebrations, and exhibitions, such as boxing champion Elio Plaisant's one, will last until May 24th.

Twenty-six days of uninterrupted activity and complex intensity, a new one for the Caudron. The plane daily routine is to fly over the town and in the skies of the lonely Pampa, regularly seen under low waters sprinkled with reeds. A lonely place with only anacondas, jaguars and savage populations; it's a forbidden and impossible landing space.

Having been conceived along La Manche shores – sea neck with sheared winds; having been tested in the fields of moors and on the sandy beaches of Le Crotoy and Le Touquet[62], all these are the main assets of this type of biplane.

Its rusticity and its shape know how to deal with the empty spaces, beaten by the heavy converging and brutal winds of the countries bordering Argentina, Uruguay, Brazil, Paraguay, Bolivia…

Mid May 1921, the G.3b already embodies evident glory, well deserved laurels, trick flying, a difficult succession of celebrity – but only for its pilot. For the plane, apart from its flights over Buenos-Aires, the surrounding districts and pampas, even inhospitable, its real tests are going to start…

°°° *The mouth of the Rio de La Plata. Sight Google Earth by Anne Vanier.*

Uruguay stands on the opposite mouth of the Rio de la Plata- zone of 230 kilometers to cross in on trial, the Rio has been for a million years this junction of the Atlantic. There, the sky above its briny and windy waters is an ideal space for real birds - these big sea birds which fly above the Parana Waters and feed there. They fish in the humid mangrove swamps which drown the landscape and fly at the same altitude as that of the over-weighted plane G.3b… In the new morning of Tuesday, May 24, 1921, the G3.b is crossing the Rio - we must note that this Caudron is a two-seater plane, like the G.3a was. And the G.3b is not yet the hydro it's going to become… so the danger is real, because with its only one petrol tank, it carries Adrienne Bolland, René Duperrier, two dogs and suitcases in the small and fragile airframe which takes off quite easily in the cool autumn air – winter will soon be arriving in South America.

The two-seater flies overweight, like three one-seater G.3... in real danger. The paths of bird and plane intersect, and this danger lasts for the two hours and twenty-five minutes of the flight over the Rio de la Plata[63].

The crew lands safely in Uruguay, where they are welcome by Angel S. Adami[64].

Montevideo is an important center on the road going to Argentina: pilots and mechanics, civilians or military, or ex militaries, came before and after the First World War.

Angel S. Adami offers friendly maintenance for the plane, and of course efficient cultural advertising for the pilot. Angel S. Adami is both a pilot and a business man; he bets on regular airlines and benefits from the support of media to communicate on the habits and traditions of Uruguay, now very much in the public eye thanks to an old Caudron piloted by a young eccentric woman.

°°° 1 – *Adrienne Bolland in a private home in Montevideo. Angel S.Adami and his wife on the right. The pioneer can be recognized with his moustache. Coll.Historical Institute of Aviation, Uruguay. 2 – Angel S. Adami welcoming Jean Mermoz in Montevideo, 1933. He becomes, with Vicente Almandos Almonacid, one of the essential men with the French Aéropostale in South America. Coll.Historical Institute of Aviation, Uruguay.*

Adrienne Bolland arrives with several official stamps on her accreditation, among which one of the « brother » flying Club of Buenos-Aires. Marcel Paillette, who's working now in Argentina, built Military Aviation in Montevideo helped by other "fans". At the same times civil aviation is taken care of by a handful of men, including Mario Garcia Cames[65], Ricardo Detomasi, and Angel Adami[66] - the creators of Atlántida, the first civilian flying club of Uruguay in 1914.

Europe is nearer to Uruguay than Argentina. We must not be surprised if the history of various sites and airfields, created by French and Italians, in South America, is the same as in France or Italy, or everywhere else. When it deals with testing planes (in the sense of "heavier than air"), a daredevil student is always ready to go. In 1910, in this far away field[67] - all the men still remember it – the "plane" flew briefly, like in everywhere else…

Until May 25th 1921, nobody had seen such a simple plane as the G.3*b* Caudron, piloted this way by a woman, a French one, whose hand is so sure and the ambition so great. The G.3*b* discovers Adrienne Bolland temperament, endless dry carpets of sand the same color as its wings, and longer beaches than Le Crotoy's in the "Baie de Somme" – that it might not have seen… Here, waters of the Atlantic are furious. Wind, salt and sand soften points and angles; little by little the ashwood of masts and the metal are disjointed. Thick mists of the jungle nearby affect the beachcombers' driftwood. Figures are sculpted by harsh wavelets. In the morning when the tide is low, the beach is an embossed patchwork.

From Wednesday May 25th to Monday June 13th 1921 : fifteen days to visit the aerial fleet of Uruguay, which has changed a lot since the Escofet 2, from Enrique Martinez Velazco, the daredevil engineering student.

The Blériot with an Anzani engine 25-hp of Armand Prévost[68] has been thrown away. As in Argentine, the "Estancieros", wealthy landowners have devoted a part of their land to their own planes taking off and landing, and these places have turned into civilian full time airfields[69].

°°° *1- The Parque Hotel, as seen from its private beach, used as a landing field in 1920. Coll. C.Béry. 2- The Parque Hotel, view from the arrival in car. Coll. C. Béry. 3 – Montevideo harbor. Coll. C.Béry. 4 – Place of Independence in Montevideo 1920. Coll. C.Béry. 5- Adrienne Bolland and Mr Frank Singleton (meteorologist, guidebooks writer and aeronaut, his pen name was Proteus). Coll. Historical Academy of Uruguayan Aviation. June 6-9 1921, meeting in Montevideo. 7 – The G.3b in the back of the shed, its pilot has just baptized of*

the South American descendants of the writer Jules Supervielle. Coll. Historical Academy of Uruguayan Aviation. 8 – In front of the G.3b MADO. propeller with four Uruguayan military pilots. Coll. Historical Academy of Uruguayan Aviation.
°°° *Advertising for the Caudron G.3 of Civil Aviation School in Uruguay, 1922.*

In Uruguay, the Blériot Type XI, the Farman Longhorn, the Avro 504K, Breguet XIV, Nieuport 27 and Castaibert 913-IV have been carefully transported by boat. Remodeled, they have to get acclimatized to the rough weather forecast: the country being beaten on its Eastern side by the huge Atlantic, and they all pass along in the air, flying low, preparing for a once-in-a-lifetime raid.

The greatest aviators[70] fly or have flown in Uruguay. Sometimes they came from France; sometimes they have been trained in France, sometime in Argentine or Brazil. Thanks to the intense traffic of the Transatlantic, and short hops, the worlds of sea, ground and air in 1921 are forged into one. All ambitious aviators dream of Rio-Montevideo-Buenos-Aires, an aeronautical feat that would make them legendary... The air pilot of the G.3*b* starts imagining a linking of those three points, in another historic raid. After the stones of the summit, the remaining Caudron is going to test ground and water.

Time flies in Uruguay and for the plane, days are busy: 9 to 11 a.m. and 3 to 5 p.m. the hours are precisely defined, and worry René Duperrier. It would be necessary to go fetch other planes in France to carry on working in South America. Then, On Monday June 13th 1921, a friend journalist pioneer of air guide books - Louise Faure-Favier - receives a telegram from Buenos-Aires: the pilot goes back to France for a few months, necessary time to receive diplomatic homage and convince Caudron to increase the technical teams. To be sure of coming back, she leaves the G.3*b* in Buenos-Aires before leaving for Bordeaux.

The G.3*b* has been put in the San Isidro French Mission where it is going to stay until September 1921.

The Four Times[71] of the Toucan

We are two months later, in Brazil; to follow the G.3*b* MADO., some data need to be changed.

The hemisphere's country is not located as far south as Argentine. And Rio de Janeiro is north of Santos harbor – the transit leading to the city of São Paulo, slightly inland. Rio is closer to Europe, especially to Portugal.

The Caudron G.3*b* MADO. has joined Adrienne Bolland in Santos - 340 kilometers south of Rio.

National press announces the « Senhorita Bolland »'s mechanic is already in Brazil, in
São Paulo, at Guapira's flying club since August 13th. Maybe René Duperrier has stayed in Buenos-Aires to watch proper loading of the plane? The pilot who had gone back to France, with animals and luggage, returns on Massilia Transatlantic. Adrienne Bolland had enough time to fly in a meeting in Brussels, with her friend Lucien Bossoutrot.

René Caudron has agreed to send her back for another exhibitions tour. The shipping of the G.3*b* MADO. to Santos is less expensive than to Rio.

From September 24th, everything changes for the G.3*b*.

Its occupants, already famous, are pursued by paparazzi. All their impressions, actions and movements are spied and commented by Brazilian press, so the pilot arrived rubbing her hands in anticipation of the work to be done: time's ready for funny and crazy things?

January 1st 1922 will be a special and important date in the history of the young Brazilian Republic. Numerous rejoicing (public and private), political, artistic, patriotic and international are forecasted. Some raids have already been precisely planned. They will be the spearheads for the centenary anniversary announced six months ago and which will last a whole year. And moreover, since the country will also welcome the International World Exhibition, all the "musts" will be gathered there during three hundred and sixty five long and marvelous days.

Art, architecture, aviation… Santos-Dumont, Fonck, Fronval and Périssé are in the program. Their aeronautical French Mission will disembark from the Massilia Transatlantic in September 1922, to be hosted in the brand new building of the replica of the small Trianon (the building will then host the French Embassy in Rio). Versailles in Rio will welcome the elite of French aviation during all the official evenings in South America.

°°° *The small Trianon, French building for the International Exhibition in Rio de Janeiro, 1922.*

Brazil's aviation has numerous links with the war office and the Board of Trade in France.

On the very civilized platform of the Ilha do Gobernador, Alberto Santos Dumont, a long time exile in France – and José Cárlos de Carvalho, have been respectively the first President of the Flying Club and its founder in Brazil. Today they are world stars. French people and every other nationality arrive at this patch of land, at salted water level of the Laguna nested in Guanabara's bay. It is a magical place in Brazil, to where aviators from around the world come to prepare for their future Raids in duos of pilot and mechanic.

In Rio, at the Ilha do Gobernador, the atmosphere is rich in loud and numerous projects, rich in various appetites and aspirations too. The president of the national flying club himself, Mauricio Lacerda, regularly meets the leaders of all the Brazilian provinces. The G.3*b* Caudron, once assembled, has no room on the Ilha do Gobernador's flying club - it was not expect, neither desired.

°°° *Rio harbor in 1920. Coll. C. Béry*
°°° *Rio, air view in 1922 and 1920. Review Aviación, coll. Historical Institute of F.A.A.*
°°° *Air view of the military school of Rio de Janeiro in 1922. Review Aviación, coll. Historical Institute of F.A.A.*

The reason is that eleven kilometers away, west at the military school of "Campos Dos Afonsos » the pupils are numerous (the airfield can be reached with tramway from midtown – "Centro do Brazil" station).

The war office has just spent five millions gold francs on brand new aeronautical material since 1920. As soon as the First World War ended, Brazil's government acquired in double quick time eighty-five planes: twenty Spad (among them ten fighters); twenty Breguet (among them ten bombers and ten scouting planes); forty-five Nieuport, and forty-five Breguet (sesquiplanes 19A). These were purchased following the advice of Colonel Seguin, aeronautics director in Rio[72]. All agreement are signed; the mechanics already enrolled.

The military sheds are bursting; the Colonel has a big project for the south of the country, wide open and dangerously bordering Uruguay. To prevent any future problem, the Colonel wants, in "Rio Grande Do Sul" Province, an imposing airfield at the center of the triangle of San Gabriel[73], Cacequi[74] and Santa Maria[75]. Put there three squadrons of fifteen planes, three squadrons of fifteen fighters and another three squadrons of bombers - all French planes – to prepare war, is better than doing nothing - or maybe not; the Colonel must fight to keep his influence with the government.

The military Naval Aviation located not far from the airfield of the Flying Club, on Euxada Island.

Its sixty-five hydroplanes are kept and maintained for three types of actions: observation, bombing and fighting: missions of "foreign" planes Curtiss, Machi and Lohner, piloted by students who realize those coming and going for a total of two hundred hours flight a year. The location hosts a technical multiple platform: aerial photo workshops, fabrication of parts and lodging of emigrating technicians. Military competition of the war office is an economic blessing and means daily action for the flying club airfield. In its determination to encourage young people into aviation, the war office subsidizes the air club of the Curtiss firm in Guapira[76], a small town north of São Paulo.

We know that the Caudron has no shed on site; the only civilian aviators to be certified are those sent by their native countries, with of course, the Brazilian pilots given preference. To maintain her celebrity, flying over the surroundings of Rio Bay should be enough to the G.3b pilot. However, the organizers feel that raids in Brazil call for something more serious than her Caudron. The minister cannot just accept a pilot with all the necessary official papers. It seems that Adrienne Bolland did not present enough official papers. Much is needed to make sure of the feat – there are more than two thousand kilometers with wild beaches beaten by wind gusts, assaulted by waves, Indians and felines – all of them man-eaters. This implies an extensive organization and fueling stations to Uruguay, and plenty of accreditations... All this must be in the international medias. Finally, no: "their" Andes are impossible to be flown by a civilian and woman pilot.

Outside São Paulo, at Campiñas[77], a desert lawless zone next to the leper hospital, the French firm Caudron has installed the civilian pilot and champion of aerobatics, Edouard Stechmann ; his school and flying club only attracts ten students a year. The Caudron will be stranded at this obscure location? Three hundred and fifty kilometers from Rio? – a kind of raid of its own, to be done any time the G.3b needs new parts. Not a logical place for Adrienne Bolland. But if she really wants, if she is really decided to realize these "Raids everywhere in Brazil" announced in the press, she can wait for her time. Nevertheless, the waiting list is long.

Meantime, Eduardo Chaves, the greatest of all Brazilian pilots, the apprenticeship companion of Roland Garros on the Blériot 25-hp, the first man to fly in the Brazilian sky, in Santos on March 8th 1912, flies in a Caudron. He will surely be able to relate her in French how his Rio-Buenos-Aires raid failed the year before.

The only interesting raid to be accomplished, which is also the most difficult for her G.3b – not an hydroplane – will be to connect the coast megalopolis of Rio de Janeiro and Buenos-Aires.

°°°° *2.000 kilometers from Rio de Janeiro to Buenos-Aires, of coasts to fly over for the 1922 Raid.*

Antonio Locatelli was the first one to try it, from Buenos-Aires, on September 8th 1919. His SVA 220-hp biplane managed to reach Porto Alegre; but in Tijuca on September 15^{th,} he crashed when landing. Plane and dream destroyed. Then a French man, his name spelled Daudt in the South American aviation magazines of the time, is not sure, but takes up the challenge. At the same date, he leaves Rio. His accident in Florianopolis stopped him. Soon after, an Englishman and a Brazilian one, Pinter and Aliatar, also take off from Rio; in a Manchi 9 Fiat engine 280-hp, but after a forced landing, the crank used to restart the engine escapes from his hand; Aliatar is hurt by the piece and falls in the water. Pinder is trying to save him when both drown in the Lagoa dos Esteves[78], at two hundred and thirty kilometers from Porto Alegre - seven hundred kilometers north of Montevideo.

The fourth trial is made in a similar plane, by lieutenant-commander, Virginius Britos de Lamare and his mechanic Silva Junior. After having left on the October 6th 1919, they want to land at Rio Grande do Sul, next to Porto Alegre on October 31st but there, too, an accident stops them. They survive.

Edu Chaves takes off in his twin-engined Caudron from Rio on October 31st. Bad weather leads him to São Paulo and an accident occurs due to engine failure, and it's the end of Chaves's raid – without injury. The sixth to try the adventure is US lieutenant Orton W. Hoover[79] ; survivor of a failure of his Curtiss's motor, followed by an accident in Santos.

Aviator Edouardo Hearne[80] and his mechanic Camilo Brezzi, are the seventh. Hearne tries it from Buenos-Aires; as a celebrity, the press watches him and the event - related in precise details by precise telegrams[81]. The pilot and his Hispano Suiza engine of 300-hp of his Bristol has all the qualities necessary for the challenge, but already they speak of a "cursed" raid: on January 29th 1920, the two men are in peril of death because of heavy rains, five hundred kilometers away from the Argentine capital. Finally they go back.

Even if accredited, the trials of Raid Buenos-Aires-Rio or Rio-Buenos-Aires are failures for all the pilots.

Edu Chaves is now in Indianopolis[82], which name suggests that the town has been taken from inhabitants of the forest by settlers – a dangerous place for white Europeans. The Indianopolis king is a milky complexion young man, aged thirty-three, flying on Curtiss, Spad and fortunately on Caudron. Chaves will help and support the G.3*b*'s pilot.

The date of Tuesday October 25th 1921, just one month before the 26th birthday of Adrienne Bolland, is well known for the G.3*h*; it is noted in the Rio de Janeiro, São Paulo and Santos newspapers : The President of the flying club has the pleasure to announce that the instructor aviators Fritz Roesler, Joao Robba, Manuel Lacerda, Franco and Amadeu da Silveira Saraiva will see the twin brother of the G.3*a*

Caudron, piloted by Senhorita Bolland, who has crossed the Andes, flying over the happy town of São Paulo. This is at the air meeting organized by the pioneer of the territory, Edu Chaves, assisted by Capitain Lafay and all the French Mission in Brazil. Posters with a smiling face of Adrienne Bolland are nailed on tree trunks.

Brazil is "wood": The Pau-Brasil[83] God is a dense, hard, ember glowing matter – its resin is used to color fabric. Trees, millions of them in the forest of Amazonia, are the wealth of the country; following the progress of settlers in the forest and the deforestation. The pau-brasil becomes timber or cabinet, or violin bows…

All this humidity, coming from the vegetal mass of pau-brasil, damages the plane of the Stechmann flying club (the one near the leper hospital). Between recurrent questions and mechanical problems, the takeoffs and the excessive fuel pressures are overwhelming. The propeller is strained, the structure stiffens; the spars creak; the engine, although preheated uses its eighty "donkeys" without being seriously considered. René Duperrier often hears "let's get away[84]!"

In December, the Caudron can be seen over Santos, over what was before a small fishing harbor, now transformed into an international economic center thanks to cotton, wood and mining industries. The "seafront with dream beaches" is a wonderful resort and exotic tropical paradise – it's totally new and has been already provided with luxury hotels.

°°° *Postcard of the Hotel - with the casino - in front of the sea, where Adrienne Bolland is installed in 1922. Coll. C.Béry.*

Their hotel is shown on a hand-painted postcard, sold near the beach. Its customers high society and international people searching luxury – like the modern comfort and marble halls. The small palace is white, colonial style; its pilasters and piers are pau-brazil wooden, with geometric and flowery design. Trees, sun and sea…Those funny naked persons in a single file with their parrot and Toucan's feathers, who prospect the sky with surprise, are Tupinamba Indians. Hunters and fishers, gatherers and formidable cannibals, indifferent to dogmas, say "civilized" inhabitants. When the Tupinamba stop watching the Caudron, to go back to their deep jungle, they keep silent.

°°° *Tupinamba Indian. Coll. C.Béry.*

We are in January 1922, already 1922. The G.3*b* MADO. has no place here: René Duperrier states that staying away from everything for a long time is also pure madness, technically speaking. The old plane, this "rafiot", is lacking maintenance; coming in and leaving Bertioga's beach – and other "sweet sandy" beaches near the jungle – isn't a good idea… It becomes a danger. All agree on the lack of profit of their Santos stay.

Watched night and day in the air, the plane and its pilot never stop and when Adrienne Bolland lands, she is very restless. The fact that she lives in a suite in this far-away palace at the end of the world, with a casino, means a lot. To get rid of her debts and bills and forget her problems, Adrienne Bolland enjoys herself, gambles at the casino – and finally wins money.

Money in her pockets and the press knows the air business re-start in Santos. The G.3*b* Caudron (forsaking raids for lack of money) is now seen on a different positive angle.

Far away and nowhere, without precise timetable and nothing else but weather forecast destinations: these are the elements of the Nirvana of any pilot "raider"[85].

Even without accreditations, Adrienne Bolland decides to visit Rio, along the coast.

She considers Santos-Rio in a hydro plane: a four-hour trip. What are four hours? – The pilot calls it a pleasure trip for the Caudron. Its twin brother has been through worse things in the Andes.

°°° *The Raid Santos-Rio, trial before the Rio-Buenos-Aires raid (2,000 kms),imagined in February 1922. View Google Earth by Anne Vanier.*
°°° *View Google Earth by Anne Vanier.*

Adding floats is part of the instruction René Duperrier received in the firm and his "permanent" training in South America with the pilot allows him to transform the G.3*b* Caudron into a hydroplane. (René Caudron has become a master in the art of building or transforming a G.3 Type into a hydro since 1913. The reputation of the firm has been built on this "hydravation,[86]" so useful along the sea. In fact, one of his ex-instructors, the very talented Auguste Maïcon, uses it to pilot on the French Riviera.)

°°° *Caudron Hydravion - mixt hydro plane. Coll. C.Béry.*
°°° *Drawing of adding floats for an amphibian Caudron G.3. (Review l'Aérophile.)*

After important transformations thanks to the money won at the casino, the plane becomes a totally new machine. Castor oil is a new material in Brazil[87] : the French have used it as a lubricant for some years, but the pilot must coax this "rototo" which "rubs[88]". Starting the engine is difficult, the slow takeoff (because of the added angle "behind" the wheels) is delicate to handle; landing, between wheels and floats, is a disaster on its first trips, fortunately not far from the beach. René Duperrier has a premonition : in case of a problem in flight, the pilot needs to set down and it would be difficult to start again, on the ocean... The only conclusion of the day is that the G.3*c* Hydro Caudron will have to include a mechanic in the raid. Two people instead of one.

Their free shows delight the journalists and the numerous tourists who made a special trip; at the beginning of February 1922 the press has not been invited to the official takeoff. The tourists can see its mechanic and pilot[89] clothed to the minimum. In the new G.3c hydro, the back part of the pilot's seat has been lightened. Without a toolbox, they take off from the ocean under the eyes of a perplexed public and after some difficulties to get out of the water; the plane stays on the crests of the waves for twenty-five minutes[90].

After a two-hour flight, when they are away from Bertioga's beach, the propeller breaks into pieces.

The G.3c falls.

°°° *Map of Indians populations in South America in 1950, and etching (17th century) of a men-eater banquet in a Tupinamba tribe.*

The two occupants are miraculously uninjured in the crash. But they are in the middle of the ocean, the Guaratuba beach is far and the pilot is a poor swimmer[91]. Her mechanic helps her by deploying the floating anchor - Sacadura Cabral[92]'s gift before they left São Paulo - so they withstand the furious backwash. The hydro is brought to shore, lifted and hung to a tree.

There is considerable damage. The propeller is broken around its bolted axis; a wing panel hangs miserable, for want of its piano strings; the engine is flooded with salt water. During the fall, the hood, the fuselage, the rigging, the pulleys, the mast and the rudders have been disintegrating: the whole plane is a disaster.

Aluminum parts can always be repaired but the cooling fin are made of steel and already attacked by corrosion. The only way is to leave on foot through the forest to reach Santos, a thirty-kilometer—walk in the jungle and the beaches, to seek repair parts, and a propeller.

°°° *Propeller of a biplane, 1920. Coll. Château de Suvigny-les-Beaune, Burgundy, France.*
°°° *Viewed in a different way, the G.3 looks like a coffin. Drawing watercolored by A.Vanier.*

The G.3's propeller is walnut-made, two and a half meters long and weighs more than thirty kilograms. Behind the two bewildered human beings, the jungle resonates with the cries of wild animals. They enter its black depth and the hydro is left alone on the beach, for more than eight days.

Being a G.3*a*, *b*, or *c*, doesn't change a thing to its fate. It remains on the sand, shaken by howling winds. Those winds bring sand and salt, night and day, on damaged floats, the plane bounces, day after day, without stopping. The Caudron now is used to scorpions, tarantulas, jaguars, shouts of howling monkeys and Tupinamba's eyes... During day time, nearby water is patrolled by brigades of sharks and orcas; they slide behind the salted rudder.

Above the level of what were rubber shock absorbers, components are salt-crusted. The engine is dead. A painful question worries René Duperrier and Adrienne Bolland: the whole French Aeronautical Mission knows they left, so why don't they come to rescue them? No answer to this question. We need to mention their accident has not been related in the papers. But this question enrages the pilot. After so much work on the machine, it survives, patched up, polished, glued, to become nearly capable of flight. Unfortunately, the petrol used to clean steel parts has emptied the tank.

°°° *It should be the « damned beach », but here it's the Trinidade beach, in Paraty, next to Rio de Janeiro. The World Factbook*

No one will ever come. This is the conclusion which is as hard as the endless wind blowing from the ocean and which prevents them from leaving. Thirst and hunger become permanent and the rip current seems impossible to handle with a rotten plane[93]. One day though, with the pilot shouting and crying, and the mechanic burning with tropical fever, the floats echo above the waves and they take off.

But soon fall again. They are on another beach, in São Sebastáo's channel, this one with footprints. The small colony is administrated by a mayor and a prefect, very welcoming. They are hosted by Emyglio Orselli in a sort of casemate opened in 1894 in a kind of town which was more the architectural result of a group of lonely but interdependent people having fled together in this forlorn place between the nothingness of the ocean and the invading jungle.

Now the pilot and her mechanic are reappearing in the columns of the papers. On March 22 1922, we learn that the engineer of number 15's telegraph post of São Paulo state, and all other states, had been warned on March 21st at 2 p.m. that the Santos Rio raid was interrupted. The colony, administrated by the prefect representing this state, hosts the famous pilot, her mechanic, and her plane "from which she never goes far[94]".

The "poor unlucky" aviatrix suffers another setback: it will oblige them to wait for the next day to take off. Surprisingly enough, the Spanish paper La Vanguardia announces two days later that they have suffered from a violent accident, but in Rio…

They are going to stay a week in São Sebastiáo without being able to remedy various problems. Trying to approach this coast full of rocks is more complicated than anticipated. The Oyapock is the only one to agree to the French SOS, in exchange for ten thousand francs. But the captain of the Guyanese boat doesn't transfer aeronautical tourists, its store-rooms are full of slaves going to Rio, and he's supposed to trade common-right prisoners on Pork Island: this is the name of this rock flooded under an oppressing vegetation starting one kilometer away from the Brazilian coast. The Ilha dos Porcos is tiny and scattered with sheet metal roofs which are huts built by perverse criminals and serial killers dedicated to die isolated on this "idyllic" island at the end of the world.

"A place of savages!" shouts the pilot when the G.3c hydro falls from the fragile ramp where all the natives are gathered and crying. The idea of being left by those famous people makes them very sad. Even the pilot is nearly overcome by this fall... The Caudron's engine and some ribs[95] are saved. On March 30th 1922, Rio is happy to welcome Arturo Sacadura Cabral and Gago Coutinho, the winners of a heroic Raid applauded by the whole world, and at Number 75 of Santo Amaro Street in Rio de Janeiro, journalists come for an interview with Adrienne Bolland - Goddess glad to be safe, but not celebrating. Her bitterness adds to resentment she has trouble to hide. For her interviewers, she becomes the "brave pilot so unlucky with her plane".

Before 1924, this quandary will not be mentioned in the press, only the successive telegrams speak of a double accident and then omit, nothing. Has Adrienne Bolland been obliged to keep silent? The fact is that this negative advertising was damageable to the diplomatic interests of the French aeronautical Mission.

°°° *75 Santo Amaro Street is in Gloria district of Rio de Janeiro. Photo Google Earth.*

René Caudron didn't respond to the pilot's requests for help and financing. By dint of Sacadura Cabral's kindness – as a survivor of this Atlantic crossing, Cabral knows what are difficulties and humiliations - the Caudron teammates – or ex-Caudron teammates, is financially saved. The papers relate the generous proposal of the year's winner for the loser: "The Navy minister, Mister Veiga Miranda, allows the damaged plane to be repaired free in Rio".

Another G.3 hydro is going to be "born", its shape and type totally unknown.

Would it be the end of the two Type G.3 hp-80 Caudron Rhône F-ABGP and F-ABGO ?

The hydro, or a heavier than air machine looking like it, or lent by one of the friends of the pilot, has been associated with the brand new Type G.3 engine Rhône 120-hp BANDEIRANTE of Anésia Pinheiro Machado. Tereza de Marzo[96] and Anésia Pinheiro has Adrienne Bolland as official witness during their Aviation License exams on Monday 8th and Tuesday 9th April 1922.

°°° *The Correio Paulistano, São Paulo's daily relates the multiple faces accident of the pilot.*

The calendar of ceremonies and aerial shows from July to the end of December 1922, which is the length of the exhibitions' program organized for the Aeronautical French Mission of Brazil, quotes the failed Raids for this new plane, and the nervous words in the bitter mouth of its pilot. All the forces present are bonded against her; even with all the joy and good temper of the world – visible on Adrienne Bolland's only official picture found at the time – she cannot face them.

Adrienne Bolland becomes the best in aerial baptisms of this part of Brazil, and from one beach to another, all winter long she initiates wealthy young girls to the pleasure of flying. She opens a "touristic air line for two people". Papers mention the name of the beaches where the public can see Adrienne Bolland land with rich clients.

Happily, no Carioca heir or government representative is in her plane the evening when the plane falls in rocks, in front of the harbor at sunset – she's with René Duperrier.

In a plane, everything goes really fast... at the moment of landing on earth and sand, two fishermen appear in a boat, invisible just before, and they have to be avoided. The sudden maneuver - but too late and too fast mixed by the fear of killing two men – have provoked a frightening visual result: that night of the end of 1922, four survivors sleep in their beds; meanwhile the plane itself is only a "wood pile".

Official end of the two Type G.3 Caudron engine Rhône 80-hp F-ABGP and F-ABGO ?

R.I.P. ? Rest in Peace, or Save Our Soul ? in Morse alphabet, S.O.S. ?

°°° *Adrienne Bolland will say later of her Brazilian experience: « 18 months of nightmare ». Here at São Paulo, flying air club during banquets of the centenary of Brazil, with Santos-Dumont, Fonck, Cabral et Coutinho.*

After this failure, these successive failures – judged as failures by the very young pilot – something much more intense is going to start. Concerning her planes, they are not sold but destroyed. The second one disappears in Rio de Janeiro in 1922, and the first one in 1921 – and the reign of the G.3*a* in Chile will cease in 1923… we know it.

The proof exists in two specimens, written with black China ink on white paper in Spanish in Chile, and in France, part of the record of registering civilian aircraft. The airplanes F-ABGP and F-ABGO are declared "lost[97]" - "dead" by the Caudron firm.

The two survivors' return is not announced by the French press. The Valdivia transatlantic brings them back to Bordeaux.

March 1923, in the sports page, we can read articles saying that "Miss Bolland and her mechanic René Duperrier are back!" and now starts the story of other planes. Planes built not only by Caudron, but by others. Anyway, all of them will offer a clear registration.

… to be continued.

End of True Birds,
now, Adrienne Bolland writes for…

Adrienne Bolland Writes For

Lecture pour Tous, published by Flammarion, in 1924.

°°° *Photos of the publication. Coll. C.Béry.*

"What are the dreams of young people? Aviation. Rising up ! Escaping ! Escape ! How many times a so-called modern literature used and misused those words, thus pretending to determine the claims of young generations, greedy of a fierce independence! It seems that present youth has reacted against this false state of mind, against this high infatuation. Sports also have largely contributed to the moral health of our young men and women. If many people dream of "escaping", it is not to live their lives away from bourgeois contingencies, but to rise up above their fellow men, and to contemplate happily the misfortunes of RATP's users, while enjoying the healthy emotions of the "joy stick". Ceiling, spin, dead stick have changed their meanings, "floating" recalls other images, and if "ungluing" for our grandmothers recalled transfers, it's no need to refine the modern meaning…

But the romantic side of aviation is not the only attractive one for young people. The primarily sportive side is, for many, the crowning of an evolution which, within twenty years, went from the bicycle (18 kilometers per hour) to the Spad (400 kilometers per hour) with the car in between. Nevertheless, aviation does not occupy, in the dreams of young people, the place of other sports; aviation is also the victim of preconceptions which it will also be able to leave, as its elders did.

The belief is that it is dangerous to climb in a plane: It is totally wrong. Another belief is that aviation needs heroism or at least an exceptional courage. Fortunately, many people get familiar with the idea of flying and piloting, and it is hard to believe how many young people, many of them young girls, question me and ask support and advice from me to "fly".

The recent trials of "moto-aviettes*" also allow-prediction of the coming popularizing of aviation. The hopes of young generations now go to the light plane, as practical and economical as the small cars. Weekends won't be spent in Deauville but in Biarritz or Nice; and honeymoons will take place around the world in less than 80 days… The faster planes will allow speed fans to replace Picardie's coast with crossing the Pyrenees and to go from Paris to Bordeaux at 180 km/hour speed; of course, I anticipate. But how much? How many months or years?

Young people's eagerness will certainly help engineers and builders to accelerate the stages and to launch aviation, if I can say, in the "public field"?"

ADRIENNE BOLLAND

* ultra light engined plane in 1924.

Time References Until the period studied in this book

1895 - November, Monday 25th, Adrienne Armande Pauline Bolland born in 2 rue des écoles, Arcueil-Cachan.
       Seventh child of Henri Bolland (Belgian) and Allonnie Pasques-Bolland (Belgian from her father and French
       from her mother).

1896 - The family moves in Paris, 114 Boulevard Arago, (district 14th).

1909 - October, Tuesday 19th, death of Henri Bolland.

1916 - Adrienne Bolland attains her majority.

1919 - November, Sunday 16th she reaches Le Crotoy, to start her pilot training at the Caudron's International
    School of aviation.

1920 - January, Monday 26th, pilot's license (FAI N°17,569) and French feminine flying club License N°12bis.
    March, she is engaged as transporter and pilot by René Caudron.
    April, Rouen's aerial show.
    August 25th, crossing of the English Channel by plane – 1st crossing by a woman from France.
    October 8,9,10, Buc's international aerial show.
    December, Saturday 4th, Adrienne Bolland, René Duperrier and two Caudron G.3 planes go to South
    America.
    December, Thursday 23rd, arrival in Buenos-Aires, Argentina.

1921 - February, Saturday 26th, she reaches 4,850 meters in a G.3..
    March, Wednesday 16th, she sends one of the two Caudrons to Mendoza.
    March, Monday 28th, 4:15 p.m., first trial of the plane.
    April, Friday 1st, she takes off at 6:20 a.m. and arrives in Santiago at 10:30 a.m.
    April, Tuesday 5th, her plane stays in Santiago.
    April, Sunday 8th, she leaves Mendoza for Buenos-Aires.
    May, Tuesday 24th, crosses the Rio de la Plata to reach Uruguay. 1st crossing by a woman.
    June, Monday 13th, comes back to Buenos Aires, leaves her plane and goes to France.
    July, Saturday 23rd, Bordeaux.
    September, Sunday 11th, returns to Rio.
    September, Saturday 24th, she's arriving in Santos in the Massilia transatlantic.
    October, Tuesday 25th, aerial show in São Paulo with the captain Lafay and Eduardo Chaves.
    December, in Santos.

1922 - January, in Santos, decides to undertake the Rio-Buenos Aires Raid.
    February, Wednesday 21st, crash in São Sebastiáo after an episode in a deserted beach.
    March, Thursday 23rd, must leave for Rio. Accident on pontoon, fall of the plane. Engine and ribs saved.
    March, Tuesday 28th, the Oyapock arrives in Rio with Adrienne Bolland and René Duperrier.
    April, Tuesday 3rd, on request of Sacadura Cabral the navy Minister (Veiga Miranda) allows her hydro plane
    to be repaired by Naval Aviation School.
    July, Tuesday 4th, aerial show in race course of São Paulo to honor Sacadura Cabral and Gago Coutinho.
    Organized by Tereza de Marzo to the benefit of Portuguese war cripples.

End of August, René Fonck, Alfred Fronval, Santos-Dumont and Yves Périssé are in charge of the
Aeronautical French Mission in Rio. They arrive on the "Massilia". Adrienne Bolland met them at the
French pavilion, then at the French Embassy.
September, Thursday 7th, beginning of the world exhibition of Rio.
October Friday 15th, opening of the air line Nichteroy-Rio.
October, Sunday 15th, opening of the air line São Francesco, in the Do Flamingo beach, trip for two persons,
(100,000 reis return ticket for each).
October, Tuesday 17th, starts passenger trips between Sacco, São Francisco, Nictheroy and Flamingo. Baptism
of Mlle Maria da Gloria Vidal, daughter of captain Christiano Jorge Vidal.
November, Tuesday 28th, flies to São Paulo with the secretary of Justice for passenger. Her project is to fly in
all Brazilian states.
Going back to Rio, accident. The second G.3 is totally destroyed.
1923 - Adrienne Bolland and René Duperrier go back to France on board the Valdivia's Transatlantic.

Thanks

and congratulations to my French, Argentine, Chilean, Spanish, Uruguayan and American friends for their help and their permanent support for six years : Gabriel Pavlovcic (former Colonel VGM (R) of the Fuerza Aerea Argentina, former director of the national Aeronautical Museum of Buenos-Aires and research worker on aviation and aeronautical history); Héctor Alarcon Carrásco (historian, research worker and writer of books on the pioneers of Chilean aviation and of history of locomotion means in Chile); Sergio Barriga Kreft (former President of the Institute of Research on Aeronautical History of Chile and Head of the Sports Aviation of the Direction at Chile's aeronautics) Fernando Puppio, pilot and editor of the Gaceta Aeronautica, in Spanish and English, on line; Gustavo Necco, doctor of meteorological sciences at the OMM (World Organization of Meteorology), advisor to the general Direction of Civil aviation in Chile in 2000, to the board of National Defense in Uruguay in 20116, research worker and writer of numerous articles on aeronautics; Sandrino Alfonso Vergara Paredes, history teacher and archivist of the Capitan Avalos Prados Aviation School of Santiago, Chile; and Dennis Simanaitis (former professor of Mathematics, Road & Track engineering editor and passionate blogger of Air themes).

Special thanks to the family of the pilot, her next of kin, her friends and to all those who, from their offices and keyboards, take part in this investigation everywhere in the world.

All my love and admiration to my marvelous son.

Sources Bibliography Websites

Archives of Boland and Sicot families.
Michel Bronstein and Margot Duhalde provides.

Femmes de l'air, Roland Tessier – Flammarion, 1948.
Princesses de l'air, Paluel Marmont – Editions G.P. Bibliothèque rouge et or, 1954.
Les grandes aviatrices conquérantes du ciel, Hervé Lauwick – Presses de la cité, 1958.
85 récits et aventures de l'air, Jeanne Beuville - Gründ, 1967.
Le temps des hélices, Général Raymond Barthélémy – Editions France-Empire, 1972.
C'est arrivé un jour, 1, Pierre Bellemare – Editions N°1, 1979.
Femmes de l'air, Marie-Josèphe de Beauregard – France Empire, 1993.
Le ciel est à elles, les premières aviatrices de Mont-de-Marsan, François Maurice – Atlantica, 2009.
La vie quotidienne dans l'aviation, en France au début du XXème siècle – Edmond Petit – Hachette, 1977.
Dictionnaire de l'aviation – Victor Houart et Edmond Petit – Editions Seghers, 1964.

Paul Deville et ses cinquante-huit prototypes – Maurice Victor – Jeanne Paul Deville éditeur, 1964.
Les frères Caudron pionniers de l'aviation – Fernand Poidevin – La Vague verte, 2009.
L'Industrie aéronautique en France 1900-1950, Emmanuel Chadeau.

Las mujeres en la aviación argentina durante el siglo XX – Gabriel Tomas Pavlovcic – Ediciones Argentinidad 2010.
La victoria de las alas – Ángel Maria Zuloaga – Edición 1958.
Ángel S. Adami: Fundador de la Aviación Civil Uruguaya - Juan Maruri.
Precursores uruguayos civiles y militares de la Aeronáutica Argentina - Juan Maruri, 2012.
Chile – Impreso por el gobierno de Chile – Empresa Zig-zag, 1915.
Ciudad y campo entre dos siglos, Buenos Aires, Cuyo y el litoral en 1890-1910 – Fotografías de Samuel Rimathé – Ediciones de la Antorcha, 2007.
Los anos del daguerrotipo – 1843-1870, primeras fotografías argentinas – Ediciones fundación Antorcha, 1995.
Efemérides, - Antonio María Biedma Recalde, 1935.
Misiones Aeronáuticas Extranjeras 1919-1924 - Julio Víctor Líroni, 1980.
Aeródromo Villa Lugano, -Eloy Martín 2013.
Godoy Fuentealba "Cóndor de los Andes" - Héctor Alarcón Carrasco 2010.

Le Patriote illustré, 11 octobre 1920.
L'illustration N°4050, 16 octobre 1920.
La presse, 11 octobre 1920 – 1921.
Le Miroir des Sports, 11 août 1921.
Les annales, N°1975, 1er mai 1921.

La Nación, 1921-1922.
La Prensa 1921-1922.
La Razón, mars 1921.
Revue Lima Victor N°8 (2005) et N°11 (2008).

El Mercurio, 1921-1961-1971.
Zig-Zag, N°842 9 avril 1921.
Los Andes. 1921

Mundo Uruguayo, 9 juin 1921.
Revista Anales, juin 1921.

O Correio Paulistano - 1920 a 1929.
O Paiz – 1920-1929.
A Razaõ – 1920-1929
O Brasil – 1920- 1935
A Gazeta – 1920- 1935
Fon Fon – 1920 -1930
Pequeno Jornal – 1920-1935

New York Times, 16 Mai 1921.
New York Tribune, 2 avril 1921 .
Chicago Tribune, avril 1921.

http://www.avionslegendaires.net/dossier/systemes-de-designation/limmatriculation-aeronefs-civils-monde/
http://aama.museeairespace.fr/amismuseeair/pages/caudron_giii_et_giv_j27ai_visite.html
http://www.janinetissot.fdaf.org/jt_caudron.htm
http://a-molet.wifeo.com/histoire-immigration.php
http://rcm45.com/raf-se5-a.html

Thank you for having chosen "True Birds, Searching for Adrienne Bolland's Two Legendary Planes", ebook by Coline Béry.

Adrienne Bolland is the first woman pilot to have crossed the Andes by plane, on the first of April 1921. But this achievement is only the most characteristic of her flamboyant crossing of the 20th Century.

Her outstanding character and her fate of committed human consciousness, fiercely amoral and rejecting all dogmas, are the opposite of all stereotypes which deliver her just "atypical" in the few books where she is mentioned (always with mistakes).

The blurred, approximations and misunderstandings appearing concerning her, have compelled the writer to deep investigation, so that they should not be repeated in the future. The research started six years ago is going on, and several publications are forecasted in 2016/2017, including a complete chronology of her life, and all public publications dealing with the theme "Adrienne Bolland".

Graphics by Anne Vanier, also visual designer for happenings: *"Poisson d'Avril On The Tarmac"* (aerial and historical trips, written by Coline Béry for aerodromes and flying clubs about Adrienne Bolland).

The illustrated version's, PDF printable 88 pages, shows 111 unpublished photos of Adrienne Bolland, her planes, documents and press articles of the period.

To order this one, please write: corderaide@orange.fr or see https://collectioncorderaide.com

"True Birds" is also published in Spanish and French.

El Nombre De Los Pájaros.

http://www.amazon.fr/gp/product/B019RXBKWC?keywords=aviacion&qid=1450966837&ref_=sr_1_11&s=english-books&sr=1-11

Le Matricule Des Oiseaux

http://www.amazon.fr/gp/product/B018O1Q71W?*Version*=1&*entries*=0

The Collection Corde Raide wishes you a nice moment reading it.

Also By Coline Béry

Marions-les, éditions Terre de Brume, 2014 – black short stories. Pages 41-53.
1er national award of the « black short story » of the Lamballe's Festival "Black on the city".

Adrienne Bolland ou les ailes de la liberté, éditions Le Passeur, 2016.

Les Lettres de la Religieuse Portugaise, Gabriel de Guilleragues - 2014 - Introduction and afterword for the French ebook by the Collection Corde Raide.
Paris Croque-Mort, Charles Virmaitre - 2015 - Introduction and afterword for the French eBook by the Collection Corde Raide.
Les Flagellants et les Flagellés, Charles Virmaitre - 2015 - Introduction and afterword for the French eBook by the Collection Corde Raide.
Aux îles de lumière, Renée Hamon, 2015 - Introduction and afterword for the French eBook by the Collection Corde Raide.
Aéropolis, Henry Kistemaeckers, 2015 - Introduction and afterword for the French eBook by the Collection Corde Raide.

Articles in 2A- Anciens Aérodromes (Sponsorised by French Interior, Air and War Ministers) review on line.

- http://www.anciens-aerodromes.com/?p=17058
- http://www.anciens-aerodromes.com/?p=17554

Item in Gaceta Aeronaútica

- "La Cordillera de los Andes y Adrienne Bolland en 1921"

Proofreading and marginal notes for 2A-Anciens Aérodromes

- http://www.anciens-aerodromes.com/?p=18116 (article by Gustavo V. Necco Carlomagno).

Coline Béry

https://www.faceBook.com/pages/Coline-B%C3%A9ry/812269085483734
https://www.faceBook.com/colinebery.auteur
http://www.aerostories.org/~aeroforums/annonces/aff.php?nummsg=2263
https://colinebery.wordpress.com/

Collection Corde Raide

https://www.faceBook.com/pages/Collection-Corde-Raide/1502610649990118

Adrienne Bolland Oficial Website (in constant construction)

https://adriennebolland.wordpress.com

Notes

[1] Adrienne Bolland's older brother was part of the crew on one of these boats, installing and monitoring telegraphic cables, in constant operation between France and Canada in 1910.
[2] Nickname of Buenos-Aires.
[3] 1hp = 746W – 1ch = 736W.
[4] Evidence of Margot Duhalde in Santiago de Chile in 2011.
[5] GPS Points : 49°23'08.03''N –1°10'35.72''E
6 Dixit Adrienne Bolland.
[7] This quoting, badly translated into French at the time, has created a diplomatic incident : The French government did not answer, thinking they needed to send young girls by liners to Chile, and Chile, annoyed by French silence, took some month to understand.
[8] Civil Aircraft Register.
[9] Finding the registration of the first planes is very difficult. Their "codes" have become secrets kept by those who have found them.

[10] GPS Points : 50°16'17.22''N – 1°40'28.90''E
[11] GPS Points : 48°49'32.13''N – 2°16'04.58''E
[12] Word used by Adrienne Bolland to name her Caudron's « Chicken cage ».
[13] G. for Gaston (1882-1915), older brother of René.
[14] The Hexagon : an affectionate nickname for beloved France; the country has this shape – vous ne trouvez pas ?
[15] In 1920, it is mentioned « firmes d'avionneurs », not yet « Planes Production ».
[16] The Caudron firm must pay the tax on war benefits, so they sell very cheaply] all excess war production. « Caudron never recovered from the return of peace. The experience acquired with the war twin-engined G.4 of 1915 and R.11 or R.23 of 1918, then his financial participations in the Aerial shipping line, and also in the Franco-Roumaine airline, prompted René Caudron to mainly develop freight planes. Caudron's industry survived thanks to small one-engine two-seater plane for training, the C.59. 250 planes from 1923 to 1926, which was a business of 11 million. This small market condemned the firm to a technical mediocrity meanwhile it saved him from financial disaster… ». *L'Industrie aéronautique en France 1900-1950*, Emmanuel Chadeau.
[17] Adrienne Bolland already was on front pages in the Argentina press for crossing the English Channel on August 25th 1920.
In some papers, Adrienne Bolland appears with her father's same – Boland – with one L; name chosen to avoid French prisons (… I'll soon write AB's family story).
[18] Pale yellow is the result of a military agreement between the countries in war during the First World War. Pale blue, sky colour and thus invisible from far away, was forbidden and reintroduced only after the conflict.
[19] The Lutetia, of the South Atlantic Company : on October 26th 1920, the Ouest Eclair newspaper published an article on the revival of fast trips between France and South America, this « thanks to the big Lutetia liner of the Compagnie Sud Atlantique : The ship took the open sea under the new postal agreement adopted by the Parliament when she left for holidays " characteristics of the ship : 15,600 tons, 175 meters length, 21 knots speed ; 4 propellers ; 345 crew people, about 1,000 passengers."
[20] Harriet Quimby's in 1912, fly over the Channel, but Quimby did it from England, to France.
[21] Ruth Law, loop the loop in 1915, https://en.wikipedia.org/wiki/Ruth_Law_Oliver.
[22] The « Petit Parisien » Monday January 3rd 1921.
[23] In France, the politics about aviation promotion has been called «Air Propaganda » until 1939. At this date, the nazis – then openly malevolent since at war against France, have given this word its most negative connotation.
[24] The French Mission and the River Plate Aviation Co. have been forced to move from the military grounds ; El Palomar, State ownership, located in the federal capital of Buenos-Aires's province and created after the decree of January 18th 1916 to the purpose of promoting the development of both civilian and military aviation. The newly called San Isidro, following the vote of the

first article of the initial decree (stipulating "it is not allowed to realize flights and landings in the limits of the military airfield in case of commercial flights meaning money involved, even if they are outside military jurisdiction") allows, at the arrival of another "Franco Argentine Company of Air Transportation" on top of that, a commercial one, Handley Page Ltd. (J.Alberto Hinds, major M.Hodgson, and Lieutenant Harry Hussey), the creation of a bigger airfield a more important one and more fitted to the development of commercial aviation.

[25] Major Shirley George Kingsley is the first to locate his DH4A (built by Geoffrey de Havilland) at the Palomar Air Club. He opens the River Plate Aviation Co there, with the help of Argentine assets and those of English residents in Argentina. Four D.H.6 for the piloting school and one D.H.16 for passenger's transportation will be bought and located in four Bessonneau (wood and textile sheds) also sheltering four Farman F.50. They are next to the Curtiss JN of Curtiss Aeroplane Export Co., welcomed by Lawrence Leon on the location.

[26] Subsecretary to Aeronautics and Air Transporation in 1919.

[27] In 1913, the Chinese government – thanks to one of their first pupils, Lieutenant Paul Bon – bought twelve biplanes from the Caudron brothers. The first planes in China.

[28] Julio Victor Lironi, in the publication « Misiones Aeronáuticas extranjeras – 1919-1924. Instituto Argentíno de Historia Aeronáutica Jorge Newbery Buenos-Aires 1980, page 111 » mentions « Níza » rather than Nice. A Spanish translation of the town in the South of France. In 1919, Auguste Maïcon creates the first important South of France air companies, the "Compagnie Méditerranéenne de Transports Aériens", in the Airfield "la Californie", where the French pioneer Ferdinand Ferber had begun some years before.

[29] Thus, Benjamin Duhau, powerful « Estanciero » (land owner) of Buenos-Aires State, has a landing field established, buys a Spad brought by Jean Guichard from France end of 1920 and asks Marcel de Saint-Sulpice to take him to his land to be taught by him in his "Estancia". This air field will be called « Estación Duhau ».

[30] Marcel Paillette, born in 1884, pilot license on June 10th 1910. At the Havre Meeting from August 25th to September 6th 1910, aboard a Blériot, he gave exhibition flights in Argentina, Chile and Paraguay. Meanwhile, he takes part in the creation of flying schools in those countries. http://www.papybleriot.fr/archives/2012/10/12/25314039.html.

[31] Since the beginning, Adrienne Bolland, during interviews, names all her planes « my » Caudron.

[32] On April 13th 1918, Lieutenant Luis Candelaria, an Argentine, crosses the South Andes in a Morane Saulnier type L « Parasol » 80-hp, on 39th parallel., where mountains are only 4,000m. On December 12th 1918, Capitain Dagoberto Godoy, Chilean, in a Bristol M.1C N° 4988 110- hp - Lo Espejo to Mendoza. He lands in Lagunita before leaving. On April 5th 1919, Capitain Armando Cortinez Mujica, Chilean, Bristol M.1C N°4987 110-hp. Lo Espejo-Tamarindos. July 30th 1919, Lieutenant Antonio Locatelli, Italian, biplane SVA Ansaldi, 220-hp, from BA to Mendoza and Santiago to Buenos-Aires. On February 10th 1920, Fernando Prieur, French, biplane Breguet Renault 300-hp,

Mendoza-Santiago-Ovalle. Drove out of course North by an error of piloting. On March 9th 1920, Antonio Parodi, Argentine, biplane SVA Ansaldi 220-hp. On March 16th 1920, Pedro Zanni, Argentine, biplane SVA Ansaldi 220-hp. On march 29th 1920, Vincente Almandos Almonacid, Argentine, biplane Spad Hispano-Suiza 220-hp, Mendoza-Vina del mar. On April 1st 1921, Adrienne Bolland, French, Mendoza-Santiago. Sesquiplane Caudron engine Rhône 80-hp. On May 22th 1921, Roberto Herrera and Alfredo Gertner, Chileans, biplane DH-22, 230-hp. Lo Espejo-San Luis. First crossing with passenger.

[33] Antonio Locatelli (1895-1936) learnt piloting in Bergamo, near the Alps. Private pilot of Gabriele d'Annunzio, who named him « the lion guard » during war. He dies in Somalia during an air attack in 1936. Gabriele d'Annunzio had a cenotaph built as an homage to his late friend.

[34] http://www.meteo-paris.com/chronique/annee/1920

[35] Adrienne Bolland notes she has to lean out from the fuselage to see clearly her position.

[36] *Les femmes de l'Air*, Roland Tessier.

[37] Revelation made to Robert Marchand on April 1st 1951, when interviewed by a Loiret paper "Ce matin-Le Pays".

[38] O Correio Paulistano March 20th 1921.

[39] The Transandine train of the Pacific railway travels a straight line from the Atlantic coast, leaving Retiro Station, not far from the mouth of the Rio de la Plata in Buenos-Aires to the Pacific coast reaching Valparaiso, Chile, thus crossing the Andes through a five kilometers tunnel dug in 1894.

[40] Ernesto Bavio, reporter sent by the paper "La Nación" for the event; Crochard, mechanic of the French Mission in Mendoza ; Don César Sagredo, influential man in Mendoza, and Don Alfredo Tabossi, chief of telegraph office, which depends on Transandine railway Pacifico.

[41] The military camp is named El Plumerillo since 1815, when the general San Martin organizes it there. The camp is five kilometers outside the city, to preclude women and alcohol. It was named Los Tamarindos at the beginning of 20th century until the 20's.

[42] Dixit Adrienne Bolland.

[43] From the: *Archivos de la linea del Ferrocarril Pacifico*, General Archives of Mendoza.

[44] Will she do it ? or not ?– "¿ Cruzará ? ¿ No cruzará ?" – The Andes paper, 1921.

[45] Duperrier, Crochard, some soldiers, the telegraph director and a journalist.

[46] The time tables are the ones of the railway coupled with the ones of the telegraph.

[47] Definition of the rudder bar by his creator, Henry Farman : "… at the end of a board fixed on the floor with a bolt – but the bolt must have play – in this axis, we connect wires; which wires are connected to the back flexible control. This mechanism makes possible to keep the good altitude in flat turning, to slide without turn over." Extract from *"La Vie Quotidienne de l'Aviation."*

[48] Publication of April 9th Buenos-Aires, translation of "… un vrai ballet pour mon Caudron, et j'ai commencé là, à jouer avec le palonnier…"

[49] GPS Points: 32°49'09.14'' S – 69°56'23.51''O.
[50] The Venturi Effect is the reduction in fluid pressure that results when a fluid flows through a constricted section (or choke) of a pipe. It is named after Giovanni Battista Venturi, Italian physicist (1746-1822).
[51] Estimated 3,9 million people in 1921.
https://es.wikipedia.org/wiki/Elecciones_parlamentarias_de_Chile_de_1921-
At the same period, Paris has 1 million less inhabitants.
[52] January 8th 1913 http://www.escueladeaviacion.cl/historia.php
[53] Arturo Alessandri Palma, lawyer, liberal party, president of the Republic twice (March 23rd 1920 to September 24th 1924 ; then March 20th 1925 to October 1st 1925). Born in Longaví, Linares province on December 20th 1868, death in Santiago du Chili on August 24th 1950.
[54] List established for the "Boletín informativo N°11 – 2015 - El Instituto de Investigaciones Histórico Aeronáuticas de Chile".
[55] Plane without wings named "roll" ("rouleur" in French), it allow learning of rudder and engine control.
[56] *"Historia de la Fuerza Aérea de Chile"*.
[57] http://ivansiminic.blogspot.fr/2012/04/sucesos-de-1920.html
[58] To know more about the internal evolution in the first flying clubs, of the first French monoplane arrived in South America in August 26th 1910, see the excellent article: http://augrsdesjours.blogspot.fr/2013/07/lhistoire-de-laviation.html, from which we quote technical parts.
[59] She nickname herself Zizi, from her younger age. She nicknames her older brother and sisters: Ben (Benoît), Mimi (Bernardine), Niche (Dieudonnée), and Doudou, one of her cousins.
[60] GPS Points : 50°13'42.27''N – 1°36'43.35''E.
[61] Friday 6th February. Adrienne Bolland is the only pilot in the world with this 12Bis number.
[62] GPS Points : Le Touquet : 50°31'29.93''N – 1°34'46.49''E
[63] The first pilot to cross the Rio de La Plata is Jorge Alejandro Newbery, with the monoplane Centenario, Blériot Rhône engine of 50-hp, on November 24 1912.
[64] Article of Gustavo Necco, historian, writer and research worker in aeronautics at Historical Institute of Uruguay, published in 2A.
[65] Mario Garcia Cames got his license in France. Angel S. Adami and Ricardo de Tomasi, in Buenos-Aires,
[66] Ibid.
[67] Paysandú.
[68] Flight of Armand Prévost on December 7th and 8th 1910 on Paysandú race course.
[69] Armand Prevost, Leopold Dolphyn, Henri Pequet, Alfred Valleton, Lowis Boyer, René Volant and some Italian pilots have met Fernando Borrell, an Uruguayan businessman, who asked them to come to Paysandú, so that the population of this town could see the flight of the first plane in this part of the country and Paysandú became an air site…

[70] Don Pablo Castaibert Jorge Newbery, Roland Garros, Juan Manuel, Boiso Lanza, Vicente Almandos Almonacid, Rufino Luro Cambaceres.

[71] Four-stroke engine. 1: intake 2: compression 3: power 4: exhaust.

[72] The French stakes in Brazil between the two wars: the military mission 1919-1940) Hugo Rogélio Suppo – *"Guerres mondiales et conflits contemporains"* No. 215 (July 2004).

[73] GPS points: 30°30'09.81''S – 54°19'28.79''O.

[74] GPS points: 29°52'32.89''S – 54°49'03.5''O.

[75] GPS points: 29°41'38.08''S – 53°49'12.34''O.

[76] GPS points: 23°29'43.37''S – 46°34'59.20''S.

[77] GPS points: 23°00'35.16''S – 47°08'18.66''O.

[78] GPS points: 28°51'20.56''S - 49°17'20.54''O.

[79] Former instructor of the naval Air School of Rio de Janeiro.

[80] Pilot license 145 on July 30th 1920 on Avro 504 in El Palomar.

[81] O Paiz, January 30th 1921.

[82] GPS points: 23°19'59.51''S – 51°08'01.03''O.

[83] "The legend says that the Portuguese settlers, disembarking in the country in 1500, met Indians shouting "Brazil !" while dancing around a tree. They thought it was the name of a plant. Error. Brazil was the name of a spirit of the forest that the Indians called forth and adored. The original name of the tree is "ibira pitanga", which means red wood in tupi guarani language."
http://www.lepetitjournal.com/sao-paulo/societe/15944-histoire--grandeur-et-ddence-du-pau-brasil.

[84] A favorite phrase of Adrienne Bolland.

[85] Pilot of air raids.

[86] The Caudron floats are covered with an external liner coated with collodion (a product well known to photographers) which is obtained by a cold solution of alcohol and ether. This device has a disadvantage: the cracks and, after, the leaks are difficult to detect. All the floats, then, are divided. They are provided with big cork screws to help empty, repair dry or dry them. Caudron and Nieuport also equipped their floats with a step, which makes the lifting off water easier. The observers note that the planes with steps floats seem to receive less violent shocks from the waves than unequipped ones. Now the floats are equipped with a shock absorber, which prevents the ruptures of supports, linking them to the air frame on the back of the Caudron, an extra light float is protected from tearing up by a "spatula" shape. Source: L'Aérophile 1913. On the Caudron, the telescopic floats are fixed on the back of the forward floats but on the Henry and Maurice Farman's hydro plane, they are also fixed on the front. The skids are attached to the floats by a mobile articulation. The back float is extra light and its only use is to prevent the airframe from disappearing in the water. To avoid the jolts of sprays, it is protected by a spatula. The floats of the J Types and K 1913 Caudron have another particularity: they contain a wheel, which freely moves in an insulated notch created inside the float. More than 130 kgs are thus gained in relation with the amphibian version. This device allows the Caudron hydro planes to take off from sandy beaches on their wheels.

[87] Article published in O Correio Paulistano on March 22th 1922.
[88] *Le Temps des Hélices,* général Barthélémy.
[89] Dogs are sent by boat to Rio.
[90] Correio Paulistano on March 29th 1922.
[91] *Le Temps des Hélices.*
[92] March 13th 1922, Gago Coutinho and Sacadura Cabral, Portuguese captains start their transatlantic crossing between Lisbonne and Brazil in a Fairey IIIC. They arrive on June 16th in the Fairey IIID named "Santa Cruz", their third plane. The two first planes have been destroyed.
[93] *Femmes de l'Air*, Paluel-Marmont.
[94] « … A aviadora francesa senhorita Bolland o seu aparelho, do qual naõ se separa". Correio Paulistano, March 22th 1922.
[95] Ash tree lattes which formed the wings, and the fabric which covered them.
[96] Tereza de Marzo (1903-1986) License N°76, April 8th 1922, and Anésia Pinheiro Machado (1904-1999) License N°77, on April 9th 1922.
[97] Air Civil Register.

https://collectioncorderaide.com

www.ingramcontent.com/pod-product-compliance
Lightning Source LLC
Chambersburg PA
CBHW031543210526
45464CB00003B/1118